Success in Science

BASIC CHEMISTSRY

GLOBE FEARON EDUCATIONAL PUBLISHER
A Division of Simon & Schuster
Upper Saddle River, New Jersey

Executive Editor: Barbara Levadi
Senior Editor: Francie Holder
Project Editors: Karen Bernhaut, Douglas Falk, Amy Jolin
Editorial Assistant: Kris Shepos-Salvatore
Editorial Development: WordWise, Inc.
Production Director: Penny Gibson
Production Editor: Walter Niedner
Interior Design and Electronic Page Production: Pencil Point Studio
Marketing Manager: Sandra Hutchison
Cover Design: Leslie Baker, Pat Smythe

Printed in the United States of America

1 2 3 4 5 6 7 8 9 10   99 98 97 96 95

ISBN 0-835-91195-0

GLOBE FEARON EDUCATIONAL PUBLISHER
A Division of Simon & Schuster
Upper Saddle River, New Jersey

Contents

Contents

Introduction to Matter and Energy

You've probably seen plenty of sci-fi television shows or movies. Have you ever seen someone "teleported" to a different place? Usually the person (or more scientifically, the person's matter) is turned into energy particles. When the energy particles arrive at the right location, they are transformed back into the matter of the person.

Changing matter to energy and back again is still a science fiction subject. But life in our universe really is based on these two things: matter and energy. Scientists study them, their properties and the changes they undergo. By learning as much as we can about them, we can find new ways to use matter and energy in our lives.

We're not quite to the point of using our knowledge of matter and energy to teleport people or objects. But perhaps you'll be the person who discovers the magical link between matter and energy that can send you to a different place in the world, or the universe, in just a few seconds.

Describing Matter

Key Words

matter:	anything that has mass and takes up space
properties:	characteristics used to describe matter
physical property:	property such as color, size, and texture that can be observed about matter while it is not reacting with anything else
chemical property:	property that describes how a substance reacts with another substance
physical change:	kind of change in which matter may look or behave differently but no new substance is formed
chemical change:	kind of change in matter in which new substances are formed

KEY IDEAS

All matter has properties that can be described. Physical properties are observed without changing the substance. Chemical properties describe how a substance changes into new substances. In a physical change, the form of the substance changes, but no new substances form. In a chemical change, new substances are produced from the original substance.

The metals and plastics used in bicycles are lightweight and strong. Scientists have figured out which substances have properties useful for the different types of cycling, such as track racing, mountain climbing, or riding on the roads around your home. These useful properties have allowed cyclists to go faster and farther than ever before.

Properties of Matter. The study of chemistry is actually the study of matter. **Matter** (MAT-ter) is anything that has mass and takes up space. Chemists try to find out what matter is made of, how matter acts, and how matter changes.

To describe any piece of matter, you can talk about its color, shape, or texture. You can also describe how it changes. All matter has **properties**, (PRAHP-uhr-teez) or characteristics that can be observed. For instance, the case of a video game system is usually made of plastic, one type of matter. Some properties of this plastic could be its color and its light weight.

The properties of one type of matter help you distinguish it from another type. Compare the plastic case of a video game system with the wood on the outside of a piano. They look different because the matter in them has different properties. Size and color are two of their different properties.

Matter has two kinds of properties. A **physical property** (FIHZ-ih-kuhl PRAHP-uhr-tee) is one that you can observe while the matter stays the same, that is, while it is not reacting with anything else. The color, weight, and size of a video game system and a piano are physical properties. These properties can be observed when no changes are taking place in the matter.

A **chemical property** (KEHM-ih-kuhl) is a property that describes how a substance reacts with, or acts with, another substance. Suppose by accident you spilled some bleach on the case of the video game system. The bleach would just form a puddle. The fact that the plastic does not react with the bleach to form another substance is a chemical property of the plastic. But suppose you spill bleach on your blue jeans. The bleach reacts with the dye in the fabric and changes its color, a property.

 1. **What is the difference between a physical and a chemical property?**

_____

_____

_____

Types of Changes. Two types of changes take place in matter. When a substance has a **physical change**, it may look or behave differently, but it is still the same substance. No new substances are formed. For example,

Fig. 1-1

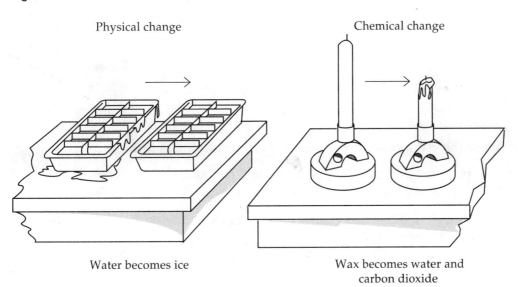

Physical change

Water becomes ice

Chemical change

Wax becomes water and carbon dioxide

water placed in a freezer hardens to form ice. See Fig. 1-1. The ice has a different size, shape, and texture than the water, but the ice and the water are the same substance. The water has undergone a physical change.

Another type of change takes place in a burning candle. A **chemical change** is one in which new substances with different chemical properties are produced. When a candle is lighted, the wick burns. The wax burns to form two gases that go into the air. See Fig. 1-1. Both the wick and the wax of the burning candle undergo a chemical change.

 2. Describe the difference between a physical and a chemical change.

_____

_____

_____

TAKE ANOTHER LOOK

Fig. 1-2 summarizes the properties and changes of matter.

Fig. 1-2

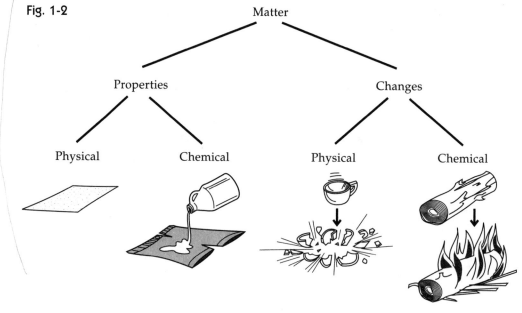

Matter

Properties

Changes

Physical   Chemical   Physical   Chemical

Check Your Understanding

Write the letter of the term in the blank next to its definition.

a. physical property   c. physical change
b. chemical property   d. chemical change

_____ 3. A substance changes in looks or behavior, but not in chemical makeup.

_____ 4. Property of matter such as color, size, and texture

_____ 5. New substances are produced from a change in matter.

_____ 6. Property of matter that describes how matter can react with other substances

Label each change as a physical change or a chemical change.

7. Iron nail rusts in water. _____

8. Gasoline is burned in a car engine. _____

9. Paper is cut into small pieces. _____

10. Tree is chopped down. _____

11. Why is it important to describe matter? _____

_____

12. Your family car is blue. Is this a physical property or a chemical property? Explain your answer. _____

_____

13. If you leave an apple slice on the counter, it reacts with the air and turns brown. Is this a physical change or a chemical change? Explain your answer. _____

_____

_____

14. When you drop a glass on the floor, it breaks into pieces, some so small you can hardly see them. Is this a physical change or a chemical change? Explain your answer. Name a physical property of the glass.

_____

_____

_____

15. Gold does not react with water to form rust. Is this a physical property or a chemical property of the gold? Explain your answer.

_____

_____

16. Plants use carbon dioxide (a gas in the air), water, and sunlight to produce their food. Is this a physical change or a chemical change? Explain your answer. _____

_____

_____

Classifying Matter

pure substance: kind of matter in which all samples of the matter have the same properties

element: kind of pure substance that is the simplest kind of matter and cannot be broken down into other substances

compound: kind of pure substance made of two or more elements joined together chemically

mixture: material formed of substances that do not combine chemically and keep their own properties

KEY IDEAS

All matter can be divided into two groups: pure substances and mixtures. Within each group, there are many different kinds of matter. These kinds of matter have different properties that make them useful for different purposes.

Wooden baseball bats are made up of compounds of the element carbon. Today, many athletes use baseball bats made of metals that have properties of light weight and strength. By studying the properties of matter and how matter combines, you can create new products that are useful not only on the ball field but also in hospitals, in factories, on farms, and in your home.

The world around you is made of matter, from the air you breathe to the pen in your hand. But all the many different types of matter actually fit into just two groups.

Pure Substances. A **pure substance** (PYOOR SUB-stuhns) is a type of matter in which all the samples of that matter have the same properties. Use water as an example. A sample of pure water will freeze to ice when the temperature is 0 degrees Celsius—0°C (you'll learn more about the property of freezing in later lessons). Another property of water is that it boils at 100 degrees Celsius— 100°C. All samples of pure water boil at the same temperature.

By performing a chemical change, scientists can break water down into two simpler substances: hydrogen and oxygen. Hydrogen and oxygen are two elements. An **element** (EHL-uh-mihnt) is the simplest form of matter that cannot be further broken down. There are 109 elements found in matter. Scientists list them in a chart called the periodic table (see Lesson 18). Only 91 of these elements are found in nature. The remaining elements have been produced in small amounts in the laboratory.

Every element has its own set of properties that are different from the properties of all other elements. Hydrogen and oxygen have two properties in common: they are both gases and they both burn easily. But they are also different. For example, humans and other animals must breathe oxygen, but not hydrogen, in order to stay alive.

Elements can join together chemically to form a pure substance called a **compound** (KAHM-pownd). Water is the compound formed when the elements hydrogen and oxygen combine chemically. The structures of elements change when they form a compound. And the properties of the compound are different from the properties of the individual elements that form the compound. For example, the properties of water are different from the properties of either hydrogen or oxygen. You've learned that hydrogen and oxygen are gases that burn easily. However, when chemically combined, they form water, a liquid that is used to put out fires.

 1. **What is the difference between an element and a compound?**

_____

_____

_____

Mixtures. Right now you're surrounded by air, an example of the other group of matter. Scientists call air a **mixture** (MIKS-chuhr), a material formed from substances that do not combine chemically. Air contains several substances, such as oxygen, hydrogen, nitrogen, and even water as a gas. But one sample of air does not contain exactly the same amount, or kind, of substances as all other air samples. Mountain climbers know that the air at the tops of high mountains contains less oxygen gas, which people need to breathe, than the air at sea level.

Each substance in a mixture keeps its own properties. Because the substances in a mixture do not combine chemically, they can be separated easily. If you leave a mixture of salt and water (salt water) in the sun, the water will dry up (it goes off into the air as a gas). You'll then see the salt left behind. This differs from the case of the water itself. To break down the compound water into its elements, you'd have to bring about a chemical change. That is, you must rearrange the structures of the elements in the compound. Remember, elements are the simplest kind of matter and can't be broken down by a chemical change.

Fig. 2-1 summarizes the relationships among the groups of matter you have just read about.

Fig. 2-1

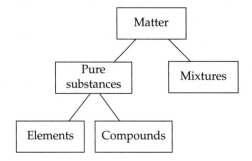

Check Your Understanding

Circle the term that correctly completes the sentence.

2. In a (pure substance / mixture), all samples of the matter have the same properties.

3. (Elements / Substances) are the simplest kind of matter.

4. (Elements / Compounds) are pure substances that cannot be broken down further.

5. (Compounds / Mixtures) can be broken down into simpler substances only by a chemical change.

6. A mixture is formed of substances that (combine / do not combine) chemically.

On the line, write whether the object is an element (E), a compound (C), or a mixture (M).

_____ 7. hydrogen

_____ 8. water

_____ 9. oxygen

_____ 10. salt water

11. What is a pure substance? _____

_____

_____

12. Are elements and compounds pure substances? Explain your answer.

_____

_____

13. Is a mixture a pure substance? Explain your answer. _____

_____

_____

14. Give an example of an element, a compound, and a mixture.

_____

_____

15. Compounds and mixtures can both be broken down into simpler substances. Explain how this happens for each type of matter.

_____

_____

_____

16. The element chlorine is a poisonous green gas. The element sodium is a poisonous silvery metal. Table salt is formed by the combination of these two elements. Is table salt a compound or a mixture? Explain your answer.

_____

_____

_____

3
Quantifying Matter

Key Words

metric system:	system of measurement based on decimal units for length, mass, and volume
meter:	standard unit of length in the metric system
mass:	amount of matter in an object
kilogram:	a unit of mass in the metric system
liter:	standard unit of volume in the metric system
volume:	amount of space, in three dimensions, that matter takes up

KEY IDEAS

One way to describe the properties of matter is with measurements. In the metric system, the standard units of length, mass, and volume are the meter, the kilogram, and the liter, respectively. Exact measurements are important in order to make good use of the properties of matter.

In preparing a prescription, a pharmacist accurately measures the various ingredients needed into a medicine bottle. The label tells the exact amount of the medicine you are to take in each dose and the number of doses you should take each day.

Exact Measurement.   A space scientist needs to figure out exactly how much fuel to store in the space shuttle. People's lives depend on this measurement. So making exact measurements is important. Similarly, sprinters must know exactly how far around the track to run for the 200-meter race.

When scientists need measurement to answer the question "how much" matter, they all use the same system. This measurement system is called SI. SI is based on the **metric system** (MEHT-rik SIHS-tuhm). The metric system is based on decimal units. Most nations of the world use the metric system. In the metric system, different units are used to measure different properties of matter. You'll see that these units are not the same as the ones you normally use in the United States, but they're easy to work with.

Length.   You often use the property of length to describe matter. In the metric system, the **meter** (m) is the standard unit used for length or distance. A person about 6 feet tall (in United States units) measures almost 2 m using SI measurements.

Mass. When you want to find out how much matter is in something, you measure the property called **mass**. Mass is different from weight. The weight of an object can change. For example, an astronaut who weighs 140 pounds on the earth would weigh only a few pounds in space. However, the astronaut's mass in both places remains the same.

Scientists want their measurements always to be the same. The mass of a person (or anything else) is always the same. For example, a mountain climbers' equipment weighs less at the top of a mountain than at the bottom. However, the mass of the equipment remains the same in each place. The standard unit of mass is the gram. One thousand grams is a **kilogram** (KIHL-uh-gram) (kg). A 110-pound person on the earth has a mass of 50 kg in SI measurement.

Now that you know the difference between weight and mass, matter can be defined more accurately as "anything that has mass and takes up space."

Volume. The **liter** (LEE-tuhr) (l) is the standard unit used to describe the property of volume in the metric system. **Volume** (VAHL-yoom) is the amount of space (in three dimensions) that matter takes up. In France, you'd buy a liter of milk, instead of a quart.

Although time is not matter, time must be measured. Think of how carefully space shuttle launches are timed. The basic unit of time in the metric system is the second (s). The other units are also the ones you're used to: minute (min) and hour (h).

 1. **Identify the metric unit used to measure the** (a) **height of a chair,** (b) **the mass of a German Shepherd, and** (c) **the volume of a gas tank.** _____

Metric Units. The basic units of the metric system can be used to describe the properties of all matter. But suppose you need to describe the volume of something small, such as, how much liquid medicine to take. You could make your measurement using the liter, but it would not be easy. The volume would be a very small number, say 0.005 l. So scientists use prefixes to indicate changes in the sizes of units. This makes the units easier to work with.

Milli- is a prefix that means $^1/_{1000}$ with whatever unit it is used. A milliliter (ml) equals $^1/_{1000}$ of a liter. Another way to say this is that there are 1000 ml in 1 l. You can use this to figure out how much medicine to take: 0.005 l × 1000 $^{ml}/_l$ = 5 ml. Using the prefix *milli-*, you know that one millimeter (mm) is one one-thousandth of a meter. And one milligram (mg) is one one-thousandth of a gram.

Another prefix is *kilo-*, which means 1000. A kilometer (km) equals 1000 times one meter; thus, there are 1000 m in 1 km. You may have seen this unit on distance signs along the highway.

Centi- is a prefix that means $1/100$. One centimeter (cm) equals $1/100$ of a meter. The centimeter is the unit often used to measure the lengths of small things. There are 100 cm in 1 m. Look at your metric ruler to see the centimeter marks.

Fig. 3-1 summarizes important units in the metric system.

Fig. 3-1

UNITS OFTEN USED IN METRIC SYSTEMS

Length
m — km (1000 m)
— cm (0.01 m)
— mm (0.001 m)

Volume
l —— ml (0.001 l)

Mass
kg — g (0.001 kg)
— mg (0.001 g)

Time
s —— ms (0.001 s)

Check Your Understanding

Write the letter of each unit in the blank next to the property of matter it describes.

a. liter   **b.** meter   **c.** second   **d.** kilogram

_____ **2.** Amount of motor oil in a can

_____ **3.** When a finish line is crossed

_____ **4.** Amount of mass in a piano

_____ **5.** Length of a piece of lumber

Circle the correct definition of each prefix used in the metric system.

6. kilo- $1/1000$ 1000

7. milli- $1/1000$ 1000

8. centi- $1/100$ 100

9. Why is matter measured? _____

_____

_____

10. Name three properties of matter that you can measure using the metric system. Then give the basic metric units used for each property.

_____

_____

_____

11. Is time a property of matter? Why is it important to measure time?

_____

_____

12. Write the shorthand form, or symbol, for each of the following units.

    (a) milliliter _____   (d) milligram _____

    (b) kilometer _____   (e) centimeter _____

    (c) kilogram _____   (f) millisecond _____

13. Arrange each group of units from smallest to largest.

    (a) km, mm, cm, m _____

    (b) mg, kg, g _____

    (c) *l*, m*l* _____

    (d) s, ms, min, h _____

14. Compare an astronaut's mass on the earth and on the moon.

_____

_____

Classifying Energy

energy:	ability to do work
potential energy:	stored energy
kinetic energy:	energy of motion
mechanical energy:	energy related to matter in motion
light energy:	energy from the sun and part of the electromagnetic spectrum
chemical energy:	energy in food, fuel, and other compounds
electric energy:	energy in an electric current
heat energy:	energy resulting from motion of particles of matter
nuclear energy:	energy in nuclei of atoms

KEY IDEAS

For every change in matter, a change in energy also takes place. Energy, or the ability to do work, can be either potential energy, which is stored energy, or kinetic energy, which is energy in motion. Energy comes in many forms. Among these are mechanical energy, light energy, chemical energy, electric energy, heat energy, and nuclear energy.

Think about some changes in matter that have taken place in your life today. Food became part of your body when you ate. Your body cells have used this food as fuel to let you move, see, and think. You probably also burned some fuel, maybe to heat your food or home, or in the engine of a car or bus while getting to school.

Potential and Kinetic Energy. All changes in matter involve changes in energy. **Energy** (EHN-uhr-jee) is the ability to do work.

Sometimes energy changes are easy to observe. One example is lighting a match. The match head bursts into flame. This chemical change gives off energy as heat and light. Other energy changes are harder to observe. For example, a small amount of energy escapes when liquid water freezes and becomes ice, a physical change. But an energy change is taking place just the same.

Picture a long line of dominoes. These dominoes have stored energy, called **potential energy** (poh-TEHN-shuhl). Thus, the dominoes have the ability to move if the conditions are right. It takes one small push to set the dominoes in motion, knocking down one after another. While falling, the dominoes have **kinetic energy** (kih-NEHT-ihk), or the energy of motion. When the dominoes again are flat on the table, they have potential energy.

 1. **Tell whether each of the following has potential energy or kinetic energy: a. a ball at the top of a hill   b. a ball rolling down a hill c. a person typing at a computer.** _____

_____

Energy can be classified into six different forms: mechanical energy, light energy, chemical energy, electric energy, heat energy, and nuclear energy.

Mechanical Energy.   You use energy to do work in an easier way. When you bicycle past a person jogging down the road, you see how much easier and faster you are moving than the jogger. The pedals and gears of the bicycle have **mechanical energy** (muh-KAN-ih-kuhl), or energy related to matter in motion. The mechanical energy of the bicycle helps you move faster and with less effort than someone on foot.

Light Energy.   Light from the sun provides the earth with most of its energy. **Light energy** is one type of electromagnetic energy, which also includes X rays and gamma rays. Plants use sunlight to make their food. The plants then become a source of energy for people, when people eat them.

Chemical Energy.   Food contains **chemical energy** (KEHM-ih-kuhl). The chemical changes that digest our food provide energy that our bodies can use. The chemical energy in food is potential energy. It has the ability to do work under the right conditions. The right conditions are digestion in our bodies. Digestion changes the chemical energy into mechanical energy that is used in moving our muscles.

Electric Energy.   Think of all the machines in your life that work when you plug them in or put in a battery. **Electric energy** (ee-LEHK-trihk) is the kind of energy in an electric current. Electric energy can often be traced back to the sun. The sun's energy causes the water on the earth to rise into the skies as a gas. When the water falls back down as rain and snow, flowing rivers form. The energy in flowing water is changed into electricity in power plants.

Heat Energy.   Heat is a type of energy, too. **Heat energy** comes from the movement of particles, or small parts, of matter. Thus it can be said that the kinetic energy of the particles results in heat. All other forms of energy can be changed into heat energy.

Nuclear Energy.   Scientists have learned that the nuclei, or inner parts, of atoms contain great amounts of energy. This is called **nuclear** (or atomic) **energy** (NOO-klee-uhr). The nuclei of some atoms can produce much more power than can be gotten from any other type of energy. In the past, the great power of nuclear energy was used in bombs that caused damage. Now, in many places, nuclear energy is used to produce electricity for everyday life.

Figure 4-1 shows relationships among various forms of energy.

Fig. 4-1

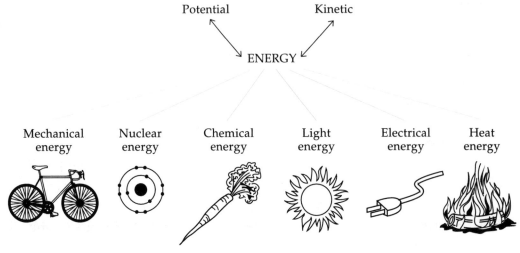

Check Your Understanding

Write a sentence explaining the connection between each pair of words.

2. potential energy, stored _____

    _____

3. kinetic energy, motion _____

    _____

4. nuclear energy, electricity _____

    _____

5. light energy, sun _____

    _____

6. heat energy, particles _____

    _____

Write the letter of each term from the list next to the phrase it describes.

a. energy
b. potential energy
c. kinetic energy
d. mechanical energy
e. chemical energy
f. nuclear energy

_____ 7. The ability to do work

_____ 8. Energy in coal, oil, and gas

_____ 9. Stored energy

_____ 10. Energy in the nuclei of atoms

_____ 11. Energy related to motion in machines

_____ 12. Energy in motion

What Do You Know?

13. What is energy? _____

14. How is potential energy different from kinetic energy?

    _____

    _____

15. Tell whether each of the following is an example of potential or kinetic energy.

    (a) child swinging on a swing _____

    (b) box of matches _____

    (c) stretched rubber band _____

    (d) carrot _____

    (e) exploding firecracker _____

16. Another type of energy is sound energy. Sound energy makes your eardrum vibrate, or move quickly back and forth. Sound energy can also break glass. Explain why sound is a type of energy.

    _____

    _____

Lesson 5
Transferring and Conserving Energy

Key Words

reaction:	chemical change in which substances combine and form new substances
exothermic reaction:	chemical change during which energy is released
endothermic reaction:	chemical change during which energy must be supplied
law of conservation of energy:	idea stating that energy can be changed from one form to another, but the total amount of energy remains the same

KEY IDEAS

Energy is given off during an exothermic reaction. Energy must be supplied during an endothermic reaction. During any chemical reaction, energy can be changed from one form to another, but the total amount of energy remains the same.

Every year on the Fourth of July, there are fireworks in cities and towns throughout the United States. People enjoy the beauty of these displays. However, fireworks can be dangerous if not handled correctly because great amounts of energy are released in these explosions.

Exothermic Reactions. Some energy changes are easy to see, like the chemical change or **reaction** (ree-AK-shuhn) that takes place when a match is lighted. The chemical energy in the match head is changed into heat and light energy. When energy is released during a chemical reaction, the reaction is called an **exothermic reaction** (ehks-oh-THER-mihk). The burning of anything is a chemical reaction that is exothermic. The energy released during an exothermic reaction was originally stored in the particles of the substances that went into the reaction. Since energy is released in an exothermic reaction, the particles of the substances produced have less energy than the original substances.

Endothermic Reactions. Some chemical reactions do not release energy. Instead, energy must be supplied to the substances during the reaction. A chemical reaction during which energy must be supplied is called an

endothermic reaction (ehn-doh-THER-mihk). An example of an endothermic reaction is when plants use light energy to produce food. The plants absorb energy to produce food that is stored in their tissues.

In an endothermic reaction, the energy supplied to the reaction is absorbed by the particles of the substances produced. Therefore, the products produced have more energy than the original substances.

 1. Explain the difference between exothermic and endothermic chemical reactions. _____

_____

_____

Energy Changes. Energy can be changed from one form to another. The sun is the source of most energy on the earth. The sun's light energy is changed into chemical energy in plants. Remains of plants and animals that were trapped inside the earth millions of years ago became fuels such as gas and oil. Such fuels can be burned in power plants to produce electricity.

Your eye is very different from a power plant, but a chemical change occurs for you to see! Light strikes special cells in your eyes, which produce electric energy. A tiny electric current travels along nerves to the brain, which interprets the image you see.

 2. What type of energy change occurs when food energy is changed to energy carried by your nerves? _____

_____

_____

Conserving Energy. Scientists have learned something about all energy changes that they have stated in a law. According to the **law of conservation of energy** (KAHN-suhr-VAY-shuhn), energy can be changed from one form to another, but the total amount of energy remains the same. That is, even though the energy may be changed into one or more different forms, the total amount of energy does not change. Another way of saying this is that energy is neither created nor destroyed during a reaction.

For example, chemists have found that when four grams of hydrogen are burned in oxygen to produce water, a precise amount of heat energy is *released*. In the reverse reaction of breaking down this water into hydrogen and oxygen by an electric current, exactly the same amount of energy is *absorbed*.

Consider a light bulb. When an electric current passes through the bulb, light and heat energy are given off. Electric energy has been changed into "useful" light energy and "unwanted" heat energy. Scientists have measured the amounts of these energies very carefully. The total amount of light and heat energy produced by the bulb is equal to the electric energy put into the bulb.

Fig. 5-1 below shows various transfers of energy.

Fig. 5-1

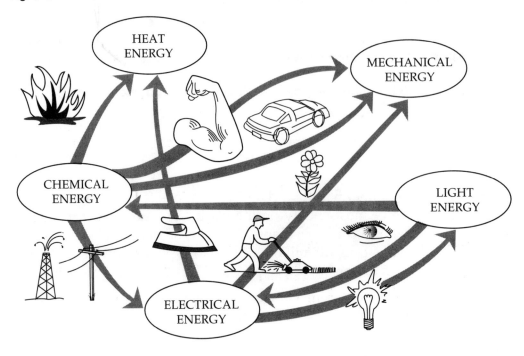

Check Your Understanding

Write the letter of the energy change that best describes each statement.

(a) chemical energy to electric energy

(b) light energy to chemical energy

(c) chemical energy to mechanical energy

(d) electric energy to heat energy

(e) electric energy to light energy

_____ 3. A toaster warms a slice of bread.

_____ 4. An airplane burns fuel when the airplane flies.

_____ 5. An apple tree produces fruit.

_____ 6. A power plant burns fuel to produce electricity.

_____ 7. You turn on the lamp to see the cat.

Circle the term that correctly completes each sentence.

8. Energy (always / never) changes during chemical reactions.

9. During an (exothermic / endothermic) reaction, energy is released.

10. During an (exothermic / endothermic) reaction, energy is supplied.

11. According to the law of conservation of energy, the total energy before and after a reaction is (always / usually) the same.

12. What is an exothermic reaction? Give an example.

   _____

   _____

13. What is an endothermic reaction? Give an example.

   _____

   _____

14. Why are exothermic reactions often very easy to observe?

   _____

   _____

15. Give an example of each of the following energy changes.

   (a) light to chemical energy _____

   _____

   (b) chemical to mechanical energy _____

   _____

   (c) electric to heat energy _____

   _____

   (d) light to electric energy _____

   _____

16. State the law of conservation of energy. _____

   _____

17. A firecracker explodes. Explain what happens to the energy released, according to the law of conservation of energy. _____

   _____

Measuring Energy Transfer

Lesson 6

Key Words

temperature:	measure of the average kinetic energy of the particles of an object
Celsius temperature scale:	temperature scale in which water freezes at 0 degrees and boils at 100 degrees
Kelvin temperature scale:	temperature scale in which the lowest reading is 0, the lowest temperature possible
absolute zero:	lowest possible temperature at which all motion stops
calorie:	amount of heat needed to raise the temperature of 1 gram of water 1 degree Celsius
joule:	unit of energy, including heat energy, in the metric system

KEY IDEAS

The temperature and heat content of matter can be measured. Temperature, which is a measure of the average kinetic energy of a substance, is measured in degrees on the Celsius and Kelvin temperature scales. The lowest possible temperature is 0 Kelvin, which is absolute zero. At this temperature, all motion stops. Heat content is measured in calories or in joules, the metric unit of energy.

Sometimes, it might get cold where you live. The South Pole is the coldest place on the earth. It is covered by miles of snow and ice. But it is still not nearly as cold as it would be at absolute zero.

Heat Energy. All forms of energy can be changed into heat energy. Heat is a measure of the motion of the particles that make up matter. Particles of matter are always moving. This is true for the particles of all matter: from your skin to the chair you are sitting on. The faster the particles move, the more heat or kinetic energy they have.

When you place a pot of water on the stove, the heat moves from the burner through the pot and into the water. The heat energy causes the kinetic energy of the water particles to increase. These particles move back and forth more and more quickly.

Measuring Temperature. To you, **temperature** (TEHM-per-uh-chuhr) means how hot or cold something is. To scientists, temperature is a measure of the *average* kinetic energy of the particles of an object. The particles of an object with a high temperature have a lot of kinetic energy.

You can measure the temperature of water as it nears its boiling point. If you place a thermometer in water that is being heated, you can see that the temperature of the water is increasing. A thermometer works because the colored liquid inside the narrow tube of the thermometer increases in volume as the temperature of the matter being measured increases. As the volume of the colored liquid increases, it rises in the thermometer. When the temperature of the matter cools, the volume of the liquid in the thermometer decreases, and the liquid falls. Temperature, of course, is measured in degrees.

Scientists use two different scales to measure temperature. The **Celsius temperature scale** (SEHL-see-uhs) is a scale divided into 100 equal units, called degrees, between the freezing point of water and the boiling point of water. The freezing point of water is 0 degrees Celsius (°C) and the boiling point of water is 100 °C.

Another temperature scale is also used by scientists. This is the **Kelvin temperature scale** (KEHL-vihn), in which the lowest reading is 0 Kelvin (K). This is the lowest possible temperature anything can reach. Zero K is **absolute zero** (AB-suh-loot ZEE-roh). At absolute zero, the kinetic energy of particles is zero. Absolute zero is also -273 °C or 273 degrees below zero on the Celsius temperature scale. Scientists have come very close to, but have not actually produced absolute zero in the laboratory. To change from Kelvin to Celsius temperature, subtract 273 from the Kelvin reading. To change from Celsius to Kelvin temperature, add 273 to the Celsius reading.

 1. **What temperature on the Celsius temperature scale is absolute zero?**_____

Measuring Heat. To the scientist, heat is a measure of the *total* kinetic energy of the particles of an object. One unit used to measure heat content is the **calorie** (KAL-uh-ree) (cal). One calorie is the amount of heat needed to raise the temperature of 1 gram of water 1 degree Celsius. The kilocalorie (kcal), which equals 1000 cal, is the amount of heat needed to raise the temperature of 1 kilogram of water 1 degree Celsius. In the metric system, the unit of energy is the **joule** (JOOL) (J). The joule is used to measure all types of energy, not just heat. One calorie is equal to about 4 joules.

2. **Define a calorie.** _____
_____

Fig. 6-1 compares three temperature scales.

Fig. 6-1

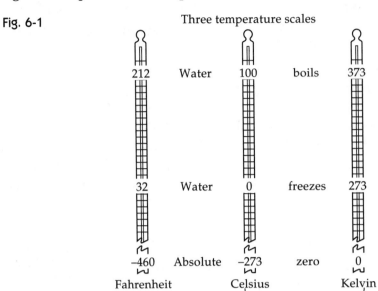

Three temperature scales

Check Your
Understanding

Write a sentence explaining the connection between each pair of terms.

3.   Celsius temperature scale, zero degrees _____

_____

4.   Kelvin temperature scale, absolute zero _____

_____

5.   -273°, absolute zero _____

_____

6.   temperature, kinetic energy _____

_____

Fill in the correct term to complete each sentence.

7.   Water freezes at _____ degrees Celsius.

8.   Water boils at _____ degrees Celsius.

9. 100 degrees Celsius equals _____ Kelvin.

10. One _____ is the amount of heat needed to raise the temperature of 1 gram of water 1 degree Celsius.

11. There are 1000 calories in 1 _____.

12. In the metric system, the unit of energy is the _____.

---

13. How are heat energy and kinetic energy related? _____ _____ _____

14. How is the Celsius temperature scale set up? _____ _____ _____

15. At what Celsius temperature does water freeze? _____

16. What is the lowest temperature on the Kelvin scale? _____

17. Explain what happens to the kinetic energy of particles at absolute zero. _____ _____

18. How is the heat of an object different from its temperature? _____ _____ _____

What Do You Know?

Summary

- Matter is anything that has mass and takes up space. The simplest form of matter is an element.

- A compound is made from two or more elements that are chemically combined. A pure substance is a single kind of matter. A mixture is a combination of elements that are not chemically joined.

- A physical property is a property such as color or size that can be observed while matter is not reacting with anything else.

- In a physical change, matter may look or behave differently, but no new substance is formed.

- A chemical property is a property that describes how one chemical substance reacts with another substance.

- In a chemical change, new substances are formed.

- The metric system is the system of measurement used by scientists. Units of length, mass and volume in the metric system are the meter, kilogram, and liter, respectively. Mass is the amount of matter in something.

- Energy is the ability to do work. Potential energy is stored energy. Kinetic energy is the energy of motion.

- Energy can be classified into six different forms: mechanical, light, chemical, electric, heat, and nuclear.

- Energy is released in an exothermic reaction. Energy must be supplied during an endothermic reaction.

- The law of conservation of energy states that energy can be changed from one form to another, but the total amount of energy remains the same.

- Temperature can be measured on the Celsius scale, in which water boils at 100 degrees and freezes at 0 degrees.

- Temperature can be measured on the Kelvin temperature scale, in which the lowest reading is zero, the lowest possible temperature. Absolute zero is the lowest possible temperature (0 Kelvin or -273 degrees Celsius), at which point all motion of particles stops.

- A calorie is the amount of heat needed to raise the temperature of 1 gram of water 1 degree Celsius.

- A joule is the measure of energy, including heat energy, in the metric system.

For Your Portfolio

1. Choose your favorite sport. Think of a piece of equipment you use in this sport that could be improved. What properties would the matter used in this new and improved equipment have?

2. Name five energy changes that have taken place in your life today.

3. Make your own pendulum. Tie an eraser or any other small weight to the end of a piece of string. Put your elbow on a table and hold the other end of the string, without the weight, in the fingers of one hand. Keep this hand still. With your other hand, pull the weight back and then let it go. Observe the motion of the pendulum. At what points does it have kinetic energy? At what points does it have potential energy? Notice about how long the pendulum swings back and forth. If air molecules didn't work against the pendulum, slowing it down, how long might the pendulum swing?

Write the letter of the term from Column II that best matches each definition in Column I.

Column I

Column II

_____ 1. combination of elements not joined chemically
_____ 2. lowest possible temperature
_____ 3. anything that has mass and takes up space
_____ 4. something that can be observed about matter
_____ 5. amount of matter
_____ 6. simplest form of matter

a. matter
b. element
c. absolute zero
d. property
e. mass
f. mixture

Fill in the blanks with the correct terms.

A(n) **(7)**_____ is a combination of elements chemically joined. The metric unit used to describe volume is the **(8)**_____. One thousand meters can also be called 1 **(9)**_____. During energy changes, the total amount of **(10)**_____ stays the same. On the Celsius scale, water boils at **(11)**_____ degrees.

Give a brief answer for the following.

12. What is the difference between a physical and a chemical change? _____
_____

13. In what units are heat and temperature measured? _____
_____

14. How is the Kelvin temperature scale different from the Celsius scale? _____
_____

Answer one of the following questions.

15. a. Define energy. Explain the difference between potential energy and kinetic energy. Give an example of each. Then list five forms of energy and give an example of each.

b. Why is it important to talk about the properties of matter? Explain why the metric system of measurement is used. What are the standard units of the metric system? How are prefixes used in this system?

Phases of Matter

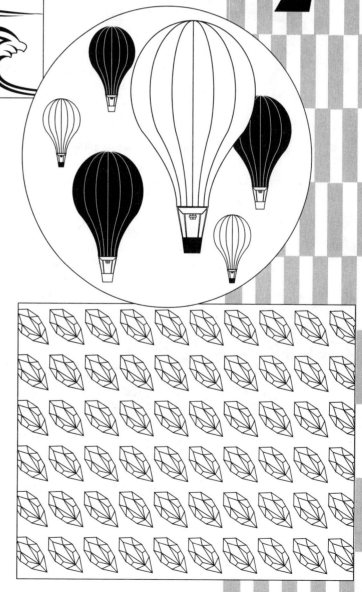

Without fresh water, people cannot survive. Most mining and industry depend on water. But people need fresh water for themselves. Each American uses an average of 75 to 300 liters of water each day. We use it in many ways—for drinking, cooking, washing dishes and clothes, flushing toilets, and watering gardens. Recently, fresh water has become a precious resource.

Clearly, fresh water is a problem in deserts and dry areas. But why is fresh water a problem in cities along the ocean? Unfortunately, people cannot drink sea water. It contains about 3.5 percent dissolved salts—and will kill you.

The island of Malta in the Mediterranean Sea is made mainly of limestone and has little underground water. In recent years, part of its water supply has come from a reverse-osmosis desalination plant along the coast. The salt is reduced from 3.5 percent to less than 0.05 percent. This level of salt is safe for drinking water.

Nature of Gases

Key Words

kinetic theory:	explanation of properties and behavior of gases
gases:	matter having no definite volume or shape
volume:	space occupied by matter
pressure:	force of a gas on the wall of a container
Kelvin temperature:	Celsius temperature plus 273°
ideal gas:	gas that obeys exactly all the gas laws

KEY IDEAS

Gases are the least dense of the three phases of matter. The molecules in a gas are much farther apart than are molecules in liquids or solids. A gas moves about freely, has no definite shape or volume, and spreads out to fill a closed container of any size.

Respiratory therapists and paramedics use oxygen gas when they work with patients. The air you breathe is about 20 percent oxygen. It is also about 79 percent nitrogen, which you cannot use. In an emergency or if you are ill, you may be given additional oxygen to breathe.

Kinetic Theory. The **kinetic theory** (kih-NEHT-ick THEE-uh-ree) describes gases. The ideas of this theory help explain the properties and behavior of **gases**. The main parts of the theory are as follows:

1. Gases are made of very small molecules.

2. Gas molecules are far apart. Most of the volume of a gas is empty space.

3. Gas molecules do not repel or attract each other.

4. Gas molecules are always moving. As they move, they bump into each other and the walls of any container they are in.

5. Temperature determines the kinetic energy of gas molecules. Therefore, gases with higher temperatures have higher kinetic energies.

 1. Most of the volume of a gas is _____

_____

Scientists usually study the properties of volume, pressure, and temperature of a gas at the same time. The **volume** (VAHL-yoom) of a gas is the total space the gas occupies inside its container. Volume units are usually given in milliliters (ml), liters (l), or cubic centimeters (cc).

Pressure (PRESH-uhr) is caused by gas molecules moving around and bumping into the sides of their container. Units of pressure include atmospheres (atm) and millimeters of mercury (mm Hg or torr). Pressure measurements are usually based upon barometric readings.

When you listen to a weather report, you usually hear the temperature and air pressure. The air pressure is measured with a barometer. In your science laboratory, you could make a barometer like the one in Fig. 7-1. In this barometer, the air presses down on the surface of the dish of mercury. As the air pushes the mercury, it is forced up into the long glass tube. Observe in the drawing that the normal or standard pressure of the atmosphere, which is 1 atm, supports a column of mercury 76 cm or 760 mm high. Thus, pressure of 1 atm = 760 mm Hg = 760 torr.

Fig. 7-1

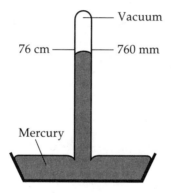

The temperature of gases is usually measured in degrees Celsius (°C) or Kelvin (K). **Kelvin temperature** (KEHL-vihn TEHM-puhr-uh-chur) is equal to the Celsius temperature plus 273 degrees.

 2. **A barometer reading shows that air pressure is 750 mm Hg. Is the air pressure above or below normal? How can you tell?** _____

_____

A gas that behaves exactly as the kinetic theory says is an **ideal gas** (eye-DEE-uhl GAS). There are no ideal gases, however, only real gases. Under ordinary conditions, real gases behave almost like ideal gases. Under conditions of low temperature and high pressure, gases do not behave as described in the kinetic theory. Hydrogen and helium are two gases that behave most like ideal gases under most conditions.

Pressure is the amount of force pushing on a unit of area. In a closed container, the pressure of a gas is caused by the molecules bumping against the sides of the container. See Fig. 7-2.

Fig. 7-2

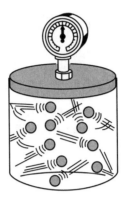

The container is closed at the top by a movable piston. By raising and lowering the piston, you can change the pressure on the gas.

Gases spread out to fill any closed container in which they are placed. See Figure 7-3. If each container has the same number of molecules, the pressure in the containers is different. The pressure of a gas is related to the average velocity of its molecules.

Fig. 7-3

A   B   C   D

Check Your Understanding

Fill in the blanks in the paragraph below.

In Fig. 7-2, if the piston was raised, the pressure would **(3)** _____.

If the piston was lowered, the distance between the molecules would

(4) _____. In Fig. 7-3, the molecules are farthest apart in container

(5) _____. The pressure is greatest in container **(6)** _____.

If the molecules in any of these containers are pushed closer together, their

velocity **(7)** _____.

The main ideas of the kinetic theory are as follows:

8. Gases are made of small _____.

9. Gas molecules do not _____ or _____ each other.

10. Gas molecules _____ and bump into each other and the walls of their container.

11. Temperature determines the _____ of the molecules.

Use the word *pressure, temperature,* or *volume* to tell which property of a gas is described in each of the following measurements.

12. _____ 730 torr

13. _____ 2 atm

14. _____ 320 K

15. _____ 87 cc

Mark each of the following True if the statement is correct or False if the statement is not correct.

16. _____ The space occupied by a gas is its volume.

17. _____ A gas has a definite volume but no definite shape.

18. _____ Real gases obey all gas laws.

19. _____ Temperatures determine the kinetic energy of molecules.

20. _____ Hydrogen is a gas that behaves almost like an ideal gas.

Behavior of Gases

Key Words

Boyle's law: volume of a gas decreases as pressure increases if temperature remains constant

Charles's law: volume of a gas increases as temperature increases if pressure remains constant

STP: standard temperature and pressure (1 atm and 273 K)

KEY IDEAS

The properties of a gas include pressure, volume, and temperature. If the size of one property of a certain amount of a gas changes, the remaining properties are affected. The size and direction of these changes can be computed if the gas is in a closed container.

SCUBA divers depend on Self Contained Underwater Breathing Apparatus for working underwater salvage. To stay under water, the diver needs a good supply of oxygen. A fairly large volume of oxygen is reduced to a small volume under high pressure in the diver's tanks. As the oxygen is released slowly from the tank, the pressure on the oxygen is reduced and the volume of the released oxygen increases.

Boyle's Law. According to **Boyle's law** (boilz law), the volume of a gas changes inversely with the pressure of the gas if the temperature remains constant. In an inverse proportion, one value increases as the other value decreases. The graph in Fig. 8-1 shows this relationship for Boyle's law when the gas is in a closed container and the temperature is constant.

Fig. 8-1

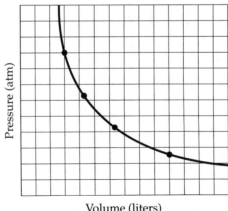

Volume (liters)

Boyle's law can be written as

$$P_1V_1 = P_2V_2$$

In this equation, P = pressure; V = volume; 1 is for the first value; and 2 is for the second value.

Problem: A 2.00-liter balloon is filled with helium gas at a pressure of 760 torr. What will be the volume of the balloon if the pressure is lowered to 608 torr?

Solution: Substitute the values into the equation for Boyle's law.

$$P_1V_1 = P_2V_2$$

$$(760 \text{ torr}) (2.00 \ l) = (608 \text{ torr}) (V_2)$$

$$V_2 = \frac{(760 \text{ torr}) (2.00 \ l)}{(608 \text{ torr})}$$

$$V_2 = 2.50 \ l$$

1. **A container with 8.00 liters of hydrogen gas has a pressure of 2.00 atm. What volume will the hydrogen occupy if the pressure is increased to 4.00 atm?**_____

Charles's Law. According to **Charles's law**, the volume of a gas increases as the Kelvin temperature increases if the pressure remains constant. The graph in Fig. 8-2 shows this relationship for Charles's law when the gas is in a closed container and the pressure is constant. Charles's law can be written as follows:

$$\frac{V_1}{T_1} = \frac{V_2}{T_2}$$

In this equation, T = temperature in Kelvin degrees; V = volume; 1 is for the first value; 2 is for the second value.

Fig. 8-2

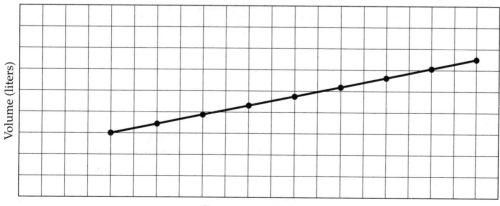

Temperature (Kelvin)

To solve the following problem, substitute the values into the equation above. Remember to always use Kelvin temperatures.

Problem: A 5.00-liter sample of argon gas at 27° C is heated to 77° C. What is the new volume of the sample?

$$\frac{V_1}{T_1} = \frac{V_2}{T_2}$$

$$\frac{5.00\ l}{300\ K} = \frac{V_2}{350\ K}$$

$$V_2 = \frac{(5.00\ l)\ (350\ K)}{(300\ K)}$$

$$V_2 = 5.83\ l$$

 2. 100 m*l* of a gas is heated from 22° C to 100° C. What is the new volume of the gas? _____

Combined Gas Laws. The volume of a gas changes when temperature (T) and pressure (P) change at the same time. You can find the new volume of the gas by combining the equations used earlier.

$$\frac{P_1V_1}{T_1} = \frac{P_2V_2}{T_2}$$

A 10.0-liter weather balloon is filled with helium at a temperature of 20° C and a pressure of 740 torr. The balloon rises in the atmosphere to a point where the temperature is -30° C and the pressure is 350 torr. What is the new volume of the balloon?

$$\frac{P_1V_1}{T_1} = \frac{P_2V_2}{T_2}$$

$$\frac{(740\ torr)\ (10.0\ l)}{293\ K} = \frac{(350\ torr)\ (V_2)}{243\ K}$$

$$V_2 = \frac{(740\ torr)\ (10.0\ l)\ (243\ K)}{(293\ K)\ (350\ torr)}$$

$$V_2 = 17.5\ l$$

 3. A 25.0-m*l* sample of carbon dioxide is at 25° C and 800 torr of pressure. To what Kelvin temperature must the gas be heated to increase its volume to 30.0 m*l* at a pressure of 900 torr?

Scientists have set conditions of temperature and pressure to have a basis for a comparison. These conditions are called **STP**, or **standard temperature and pressure**. Standard pressure is normal barometric pressure (1 atm or 760 torr), and standard temperature is the freezing temperature of water (0° C or 273 K).

See Fig. 8-3. A tank holds 3.0 liters of air at 8.0 atm and 25.0° C. How much space will the air occupy at STP?

Fig. 8-3

$$\frac{P_1V_1}{T_1} = \frac{P_2V_2}{T_2}$$

$$\frac{(8.0 \text{ atm}) (3.0 \, l)}{298 \text{ K}} = \frac{(1.0 \text{ atm}) (V_2)}{273 \text{ K}}$$

$$(V_2)(1.00 \text{ atm}) (298 \text{ k}) = (8.0 \text{ atm}) (3.0 \, l) (273 \text{ K})$$

$$V_2 = \frac{(8.0 \text{ atm}) (3.0 \, l) (273 \text{ K})}{(1.0 \text{ atm}) (298 \text{ K})}$$

$$V_2 = 22.0 \, l$$

 4. A 2.00-liter tank of medical oxygen gas has a pressure of 6.00 atm at a temperature of 30.0° C. What will the volume be at STP?

GAS LAW	PROPERTIES	EQUATION
Boyle's Law	Pressure Volume	$P_1V_1 = P_2V_2$
Charles's Law	Temperature Volume	$\dfrac{V_1}{T_1} = \dfrac{V_2}{T_2}$
Combined Law	Temperature Pressure Volume	$\dfrac{P_1V_1}{T_1} = \dfrac{P_2V_2}{T_2}$

TAKE
ANOTHER
LOOK

Heating a gas causes it to expand, but the number of molecules stays the same. As a gas expands, its volume increases. See Fig. 8-4. The rate of expansion is directly proportional to the Kelvin temperature.

Fig. 8-4

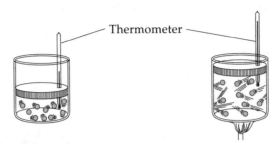

When the pressure on a gas increases, its volume decreases. See Fig. 8-5. Notice that the temperature and the number of molecules of gas stay the same.

Fig. 8-5

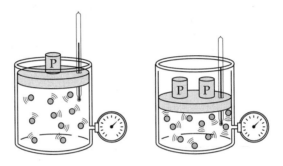

Check Your Understanding

5. In the boxes below, write the equations for the gas laws.

(a)	(b)	(c)

Charles's Law   Boyle's Law   Combined Law

Use the term *increases* or *decreases* to describe what happens when a property of a gas changes.

6. Pressure increases when volume _____.

7. Volume _____ when temperature decreases.

8. When temperature increases, volume _____.

9. Volume increases when pressure _____.

Define STP in terms of

10. degrees Celsius _____.

11. degrees Kelvin _____.

12. pressure in atmospheres _____.

13. pressure in torr _____.

14. pressure in mm _____.

Write a *B* if the statement describes Boyle's law, *C* if the statement describes Charles's law, and *STP* if the statement describes standard temperature or standard pressure.

15. _____ $P_1V_1 = P_2V_2$

16. _____ 0° C

17. _____ pressure is inversely proportional to volume

18. _____ $\dfrac{V_1}{T_1} = \dfrac{V_2}{T_2}$

19. _____ 760 torr

20. _____ 1 atm

21. _____ volume varies with Kelvin temperature

22. _____ 273 K

23. _____ volume decreases when temperature decreases

24. _____ volume increases when pressure decreases

Underline the correct answer.

25. A 100-m*l* sample of nitrogen gas is at a pressure of 380 torr. If the pressure of the gas is reduced to 190 torr, the new volume is (*200 ml,   50.0 ml,   90.0 ml,   290 ml*).

26. 30.0 m*l* of neon gas is at 273 K. The temperature is increased to 364 K. What is the new volume of the gas in m*l*? (*15.0,   40.0,   60.0,   90.0*)

27. 20.0 liters of acetylene gas at STP are forced into a 2.0-liter tank at 300 K. What is the pressure of the gas in atm? (*2.00,   11.0,   40.0,   546*)

Condensed States of Matter

Key Words

solid:	phase of matter having a definite shape and a definite volume
liquid:	phase of matter having a definite volume but taking the shape of its container
vapor:	gaseous phase of a substance that is a solid or liquid at room temperature
evaporation:	escape of vapor at the surface of a liquid
vapor pressure:	pressure of vapor above a liquid
phase change:	moving from one phase of matter to another
condensation:	changing of a gas to a liquid by cooling

KEY IDEAS

When cooled, gases condense to form liquids and solids. The forces holding together the molecules (or other liquids and solids) in a solid are strong enough to give it a rigid shape. In a liquid, the forces between molecules are midway in strength between those in solids and gases. Because of this, a liquid has a definite volume, but also flows and takes the shape of its container.

Solids. You may awake on a winter morning to find the ground covered in frost. Frost is ice crystals. This **solid** (SAHL-ihd) form of water has a definite shape and volume.

Fig. 9-1

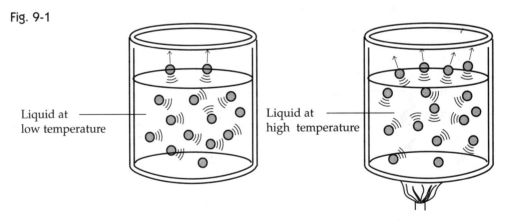

Liquid at low temperature

Liquid at high temperature

The molecules of a liquid move faster as temperature increases.

Liquids. Substances that are **liquids** (LIH-kwidz) have a definite volume but take the shape of their container. The molecules of a liquid are close together and are always moving. Not all the molecules in a liquid have the same amount of energy. Some molecules near the surface may have enough energy to escape into the air, as shown in Fig. 9-1.

The molecules of a liquid escape into the air as a **vapor** (VAY-puhr)—a gaseous state of a substance that is usually a liquid. The formation of a vapor by the escape of molecules at the surface of a liquid is **evaporation** (ee-vap-uh-RAY-shuhn). As temperature increases, the molecules move faster and overcome the forces holding them together. The amount of energy and the rate of evaporation increase. See Fig. 9-1.

The vapor above a liquid exerts **vapor pressure** (VAY-puhr PREHSH-uhr). The pressure increases as the temperature increases. When a liquid boils, the vapor pressure is great enough to make bubbles in the liquid. Fig. 9-2 shows the vapor pressure of two liquids at different temperatures. A liquid boils when its vapor pressure equals atmospheric pressure. Because liquids have specific vapor pressures, each liquid boils at its own special temperature.

The normal boiling point of water is 100° C. Note that at this temperature, the vapor pressure is 760 torr, which is normal air pressure.

Fig. 9-2

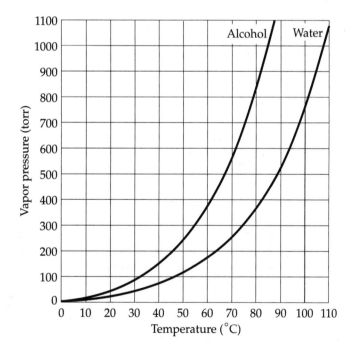

1. What is the boiling temperature of water at a pressure of 525.8 torr? _____

2. What is the boiling point of alcohol at 760 torr? _____

Phase Changes.   When heat is added to a liquid, it can change to a gas by boiling. Removing heat from a liquid can freeze it into a solid. When matter changes from one phase to another, a **phase change** (fayz) takes place. **Condensation** (kahn-duhn-SAY-shuhn) is a phase change from a gas phase to a liquid phase. Melting is a phase change from solid to liquid.

A heating-cooling curve, or phase change diagram, is used to show states of matter and how they change when heat is added or removed. Look at the curve in Fig. 9-3. As heat is added, the temperature increases until the melting point of the solid is reached. At the melting point, the temperature does not rise because energy is needed to overcome the forces that hold together the vibrating molecules of the solid.

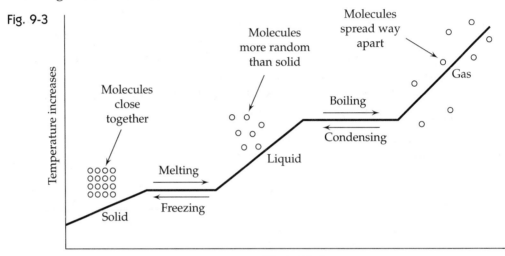

Fig. 9-3

After melting is complete, and as more energy is added, the temperature of the liquid rises until the boiling point is reached. Here energy is again needed to overcome the forces that hold the molecules together. The temperature does not rise while this is happening.

Use the table to compare the characteristics of solids, liquids, and gases.

TAKE ANOTHER LOOK

Fig. 9-4

	SOLID	LIQUID	GAS
Shape	Has its own shape	Has shape of container	Has shape of container
Volume	Definite	Definite	Fills Container
Motion of molecules	Slow (vibrating)	⟵ In between ⟶	Fast
Pattern of molecules	Definite	⟵ In between ⟶	Random
Distance between molecules	Close	⟵ In between ⟶	Far apart

Fill in the blanks in the following sentences.

3. Gases and _____ take the shape of their containers.

4. Molecules move fastest in _____.

5. There is a definite pattern of molecules in _____.

6. The distance between molecules is greatest in _____.

7. The forces holding molecules together are strongest in _____.

8. Molecules at the surface of the liquid change to a gas by _____.

9. Forces holding molecules together must be overcome before a liquid

    becomes a _____.

Write a number from Fig. 9-5 that matches each of the following terms.

Fig. 9-5

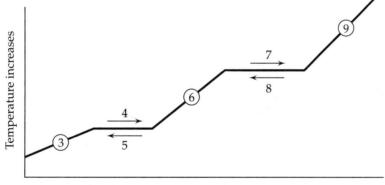

Heat added

10. _____ solid   14. _____ freezing

11. _____ liquid   15. _____ boiling

12. _____ gas   16. _____ condensing

13. _____ melting

Use Fig. 9-2 to answer the following.

17. Which liquid boils at a lower temperature, alcohol or water?

    _____

18. If a sample of water is boiling at 80° C, the vapor pressure is

    _____.

Key Words

heat of fusion:	amount of energy needed to change one gram of solid to liquid at its melting temperature
heat of vaporization:	amount of energy needed to change one gram of liquid to gas at its boiling temperature
distillation:	method of separating dissolved substances from a liquid by differences in boiling points
desalination:	process that removes salt from sea water
surface tension:	apparent "skin" effect on the surface of a liquid

KEY IDEAS

Energy is absorbed as water changes from one phase to another. While ice is at its melting temperature, a specific amount of energy is used to overcome forces between molecules. A greater amount of energy is absorbed to overcome similar forces when liquid water is at its boiling temperature. Dissolved substances are removed from water in several ways. Forces between molecules also account for the surface tension of water.

Bodies of water cover about 75 percent of the earth's surface. There is also water in the earth as groundwater, in glaciers as ice, and in the atmosphere as clouds and humidity. Water is present in all living things. About 70 percent of the human body is water.

Fig. 10-1

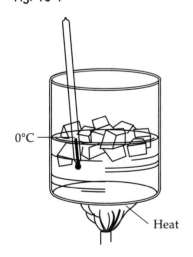

0°C

Heat

Energy and Phase Changes. The energy needed to change phases is either absorbed or given off. The amount of heat absorbed to melt one gram of solid is its **heat of fusion** (FYOO-zhuhn). The heat of fusion of solid water (ice) is 80 calories per gram. All 80 calories are used to change phases. The temperature does not increase. Look at Fig. 10-1 showing heat of fusion.

 1. **How many calories would be needed to melt 10 grams of ice at 0° C?**_____

The amount of energy absorbed to boil one gram of liquid is its **heat of vaporization** (vay-puhr-uh-ZAY-shuhn). Water has a heat of vaporization of 540 calories per gram. Look at Fig. 10-2 showing heat of vaporization.

Fig. 10-2

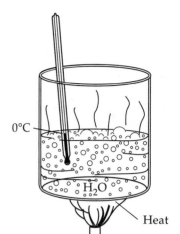

 2. **How many grams of water at 100° C could be changed from liquid to gas if 4320 calories of heat were used?**

_____

Distillation. Boiling water produces steam. Condensing the steam produces pure water. The process of boiling and condensing to separate the parts of a mixture is called **distillation** (dihs-tuh-LAY-shuhn). In the process of distillation, a flask with a liquid containing salts and other dissolved substances is heated, as shown in Fig. 10-3.

Fig. 10-3

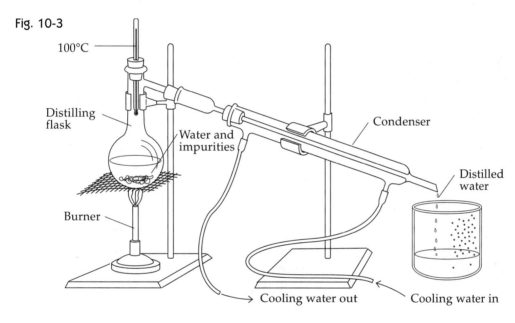

Materials in the flask that turn into gases at temperatures below 100° C, will vaporize first. When the temperature reaches 100° C, water will change to the gaseous phase. The gas will flow into the condenser. The cool water in the jacket of the condenser will absorb some of the heat. This causes the water in the condenser to return to the liquid phase. Liquid water can be seen dripping out of the condenser and into the beaker. Impurities with boiling points above 100° C will remain in the distilling flask.

Desalination. A method of obtaining fresh water from sea water is desalination. Although your body requires salt, the high percentage of salt in sea water makes it unfit for human use. The process of removing salt from sea water is **desalination** (dee-sal-uh-NAY-shuhn). Look at Fig. 10-4, which shows desalination by reverse osmosis.

Fig. 10-4

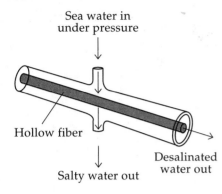

Sea water in
under pressure

Hollow fiber

Salty water out

Desalinated
water out

In reverse osmosis, sea water, under very high pressure, is forced into a tube surrounding a tiny hollow fiber. The water, but not the salt, passes into the center of the fiber. Desalinated water flows from the inside of the fiber. Each tube actually contains more than three million fibers, each about the size of a human hair.

Surface Tension. Perhaps you have floated a small needle on the surface of a glass of water. Why did the needle float? Notice in Fig. 10-5 that the water molecules on the surface of the water are only partly surrounded by other water molecules. So the molecules on the surface of the water are acted on by *unbalanced* forces. As a result, the molecules on the surface are drawn inward.

Fig. 10-5

In contrast, the molecules within the body of the water are surrounded by other water molecules. Hence the forces on these molecules are *balanced* forces. The unbalanced forces on the surface water molecules help to explain **surface tension** (SER-fuhs TEHN-shuhn), the apparent "skin" effect on the surface of the water. Because of surface tension the needle floated.

When a small amount of water is placed on waxed paper, the water beads up into tiny spheres because of surface tension. The unbalanced forces on the water's surface pull the molecules inward to form tiny beads.

In distillation, water changes phase. This process uses energy. Figure 10-6 shows the heating curve for water. At 0° C, 80 calories are used to melt one gram of ice (heat of fusion). To change liquid water to steam (heat of vaporization) at 100° C, 540 calories per gram are used.

Fig. 10-6

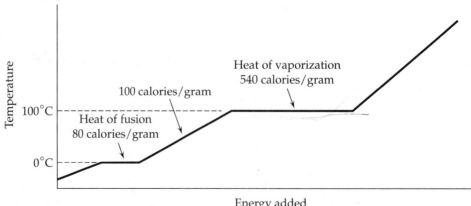

Temperature

100°C

0°C

Heat of vaporization
540 calories/gram

100 calories/gram

Heat of fusion
80 calories/gram

Energy added

3. Study the cooling diagram for water. It starts with steam at a temperature of 120° C and cools the water to -20° C. In the blanks labeled *a*, *b*, *c*, and *d*, write labels for *liquid water, solid water, heat of fusion,* and *heat of vaporization.*

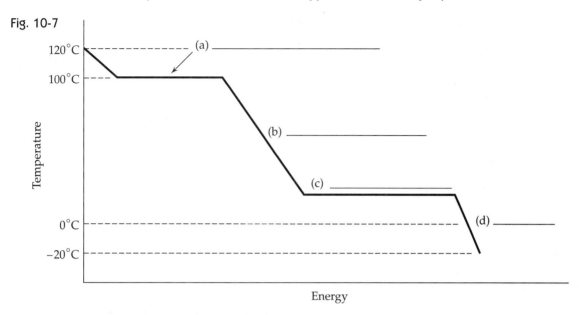

Fig. 10-7

Fill in the blanks to complete the statements.

4. The heat of fusion of water is _____ calories per gram.

5. The heat of vaporization of water is _____ calories per gram.

6. If the temperature reached only 95° C, the water remains in the _____ phase.

For each true statement write the word *True* in the space. For each false statement, write the word that, when substituted for the underlined word or phrase, makes the statement true.

7. _____ The energy needed to melt a solid is its heat of <u>vaporization</u>.

8. _____ The apparent "skin" effect on the surface of a liquid is <u>surface tension</u>.

9. _____ Removing salt from sea water is known as <u>distillation</u>.

10. _____ The changing of liquid water to steam <u>gives off</u> heat.

11. _____ <u>Distillation</u> is a method of separating dissolved substances from water by boiling and condensing.

12. _____ Steam at 100° C contains <u>more</u> heat than liquid water at 100° C.

Summary

- Gas molecules are far apart and move freely. Gases have no definite shape or volume and fill any container.

- Pressure is caused by gas molecules bumping into the walls of their container. 1 atm = 760 torr = 760 mm Hg.

- Hydrogen and helium act most like ideal gases.

- Boyle's law states that as pressure increases, the volume of a gas increases: $P_1V_1 = P_2V_2$.

- Charles's law states that as the Kelvin temperature increases, the volume of a gas increases: $V_1/T_2 = V_2/T_2$.

- The expression for the combined gas laws is $\dfrac{P_1V_1}{T_1} = \dfrac{P_2V_2}{T_2}$.

- Liquid particles are close together and in motion. Liquids have no definite shape, have a definite volume, and take the shape of their containers.

- The escape of vapor from the surface of a liquid is evaporation.

- Vapor exerts pressure. A liquid boils when its vapor pressure equals atmospheric pressure. The normal boiling temperature of water is 100° C or 273 K.

- Energy must be absorbed or given off for a substance to change phase. The heat of fusion of water is 80 calories per gram. The heat of vaporization of water is 540 calories per gram.

- The separation of dissolved substances from a liquid by boiling and condensing is distillation.

- The removal of salt from sea water is desalination.

- Surface tension is the apparent "skin" effect on the surface of a liquid.

For Your Portfolio

1. Write a speech about removing salt from sea water in dry regions of the world. Perhaps you can arrange to present the speech on cable TV local access channels.

2. Prepare a spreadsheet for the computer that will compute the costs per liter of different bottled gases at retail prices and three discounted prices.

3. A manufacturer of bottled carbon dioxide wants increased sales and has hired you to develop an advertising campaign. Design the basic ideas for this promotion.

4. Investigate and report on what happens if divers in the deep parts of the ocean ascend too rapidly to the surface.

5. Steam calliopes used in circuses operate on steam under pressure. Trace the mechanics of making music using steam power.

6. The United States maintains a reserve of helium gas. Assume you are a politician and want to continue this program but at a lower cost. Prepare your case.

7. Design a graphic labeling system for gases that will show properties such as poisonous, explosive, oxidizer, and radioactive.

Fill in the blanks in each of the following questions.

1. The pressure on 200 m*l* of oxygen gas changes from 380 mm Hg to 760mm Hg. The temperature does not change. The new volume of the oxygen gas is _____ m*l*.

2. The energy needed to change 1 gram of solid to 1 gram of liquid at its melting point is called the heat of _____.

3. The volume of a sample of carbon dioxide gas is 2 liters. As the temperature decreases, the volume _____.

4. If the volume of a gas decreases, the number of molecules of gas _____.

5. If the temperature of a liquid decreases, the vapor pressure _____.

Fill in the blank with the letter of the term that best completes each statement.

_____ 6. When water freezes, each gram of water

   a. gains 80 calories.   **c.** gains 540 calories.
   b. loses 80 calories.   **d.** loses 540 calories.

_____ 7. When the vapor pressure of a liquid equals atmospheric pressure, the liquid will

   a. freeze.   **b.** boil.   **c.** melt.   **d.** condense.

_____ 8. Of the following, the change of phase that gives off heat is

   a. solid to liquid.   **c.** gas to liquid.
   b. liquid to gas.   **d.** solid to gas.

Column I contains ideas about phases of matter. Column II contains related terms. On the line at the left of each item in Column I, write the letter of the related term in Column II.

	Column I		Column II
_____	9. volume of a gas varies with temperature	**a.**	80 calories
_____	10. volume of a gas varies with pressure	**b.**	540 calories
_____	11. volume of a gas varies with temperature and pressure	**c.**	Boyle's law
_____	12. gaseous state of water	**d.**	Charles's law
_____	13. released when one gram of water vapor becomes liquid water	**e.**	combined gas law
_____	14. absorbed when one gram of ice melts	**f.**	vapor

Write a paragraph to answer one of the following questions.

15. **(a)** Use the kinetic theory to explain Boyle's law and Charles's law. **(b)** In the heating curve for water, there are two horizontal lines. What happens to the temperature during the time represented by these lines? What happens to the energy changes? Explain.

Structure of Matter

Alchemists were people who once tried to change lead and other "base" metals into gold. In their workshops, they prepared mixtures of metals that looked like gold. Alchemists often used fraud to make their work appear successful. For example, they covered a nail that was half iron and half gold with black ink. When they dipped the nail into a liquid and stirred, the black ink came off. The part of the nail dipped in the liquid appeared to turn into gold.

Today, scientists use a giant particle accelerator to achieve the goal of the alchemist: to turn a base metal into gold. The scientists bombard mercury with the nuclei of heavy hydrogen to produce atoms of gold. Unfortunately, the gold has a short half-life. In only 2.7 days, half of the gold changes to another element. Moreover, it costs much more to change the mercury into gold than it does to buy an equal amount of gold.

Subatomic Particles

Key Words

atomic particles:	particles found inside an atom
electrons:	atomic particles with negative charge
protons:	atomic particles with positive charge
neutrons:	atomic particles with no electric charge

KEY IDEAS

Atoms are not the simplest particles of matter. Atoms are made of electrons, protons, and neutrons. Electrons and protons are charged particles. Neutrons have no charge.

In 1808, John Dalton suggested that all matter is made of atoms. Dalton thought atoms were the smallest possible particles. In the 1890s, J.J. Thomson showed that there were still smaller particles with electric charges.

Positive and Negative Charges. There are two types of charges: positive and negative. These are unlike charges. Positive charges attract negative charges. Negative charges repel negative charges. Positive charges repel positive charges. These statements can be summed up by saying unlike charges attract and like charges repel. See Figure 11-1.

Fig. 11-1

<table>
<tr><td>

LIKE CHARGES REPEL

Charges **A** and **B**, both positive, move away from each other because like charges repel.

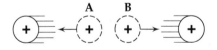

Charges **C** and **D**, both negative, move away from each other because like charges repel.

</td><td>

UNLIKE CHARGES ATTRACT

Charges **E** and **F**, one positive, the other negative, move toward each other because unlike charges attract.

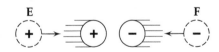

</td></tr>
</table>

Showing repulsion between like charges is easy. Stick two pieces of transparent tape to the top of a desk. Pull up the two strips. Bring one piece near the other. The strips repel each other because each picks up the same type of charge when pulled up from the desk. See Fig. 11-2.

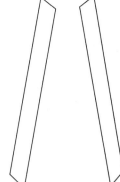

Fig. 11-2

Not only do charges attract and repel, but they also can cancel each other. One unit of positive charge cancels the effect of one unit of negative charge. Thus, a particle with one positive charge and one negative charge has no overall charge. That is, it has no net charge. What about a particle with two positive charges and one negative charge? It has a net charge of positive one.

 1. A particle has one positive charge and two negative charges. What is its net charge? _____

Charged particles. In the 1870s, William Crookes had removed most of the air from a glass tube. He found that a yellow-green glow could be made to appear at the end of his tube. See Fig. 11-3. If an object was placed midway down the tube, the object cast a shadow in the yellow-green glow. See Fig. 11-4. Crookes thought the glow was caused by radiation. He also believed the radiation came from the negatively charged plate, called the cathode. Scientists called the radiation cathode rays.

Fig. 11-3

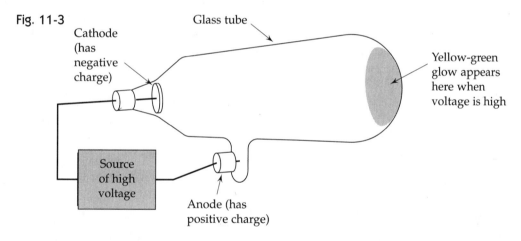

Glass tube

Cathode (has negative charge)

Yellow-green glow appears here when voltage is high

Source of high voltage

Anode (has positive charge)

Fig. 11-4

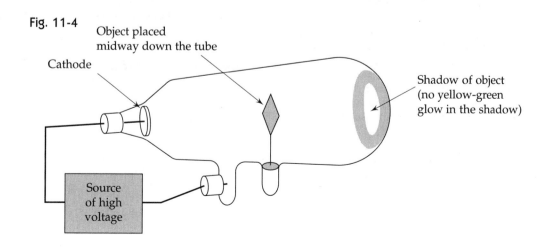

Object placed midway down the tube

Cathode

Shadow of object (no yellow-green glow in the shadow)

Source of high voltage

After repeating Crookes's work, J.J. Thomson was able to show that atoms are made up of at least two types of **atomic particles** (uh-TAHM-ihk PAHRT-ih-kuhlz). Thomson found that charged plates made the cathode rays bend. See Fig. 11-5. Thomson knew that light cannot be bent by charged plates. He therefore believed that cathode rays must be made of charged particles. He observed that the particles were attracted to the positively charged plate, called an anode. He therefore concluded that the particles had to be negatively charged.

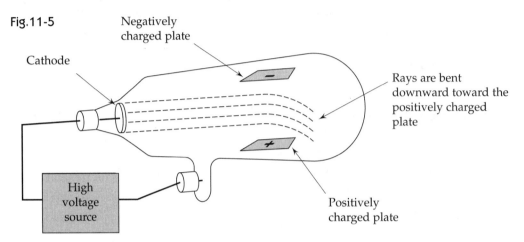

Fig.11-5

Negatively charged plate

Cathode

Rays are bent downward toward the positively charged plate

High voltage source

Positively charged plate

Thomson decided that the same charged particles must be present in the atoms of every element. These particles are called **electrons** (ee-LEHK-trahnz). The charge on an electron is the smallest possible negative charge. An electron's charge is often described as 1- (one minus).

Matter is usually neutral. Therefore, atoms must also contain particles with positive charges. The positive charges of these particles would balance the negative charges from the electrons. To search for these particles, Thomson experimented with a tube containing hydrogen gas. He found that positive particles moved away from the anode. The positive particles were attracted toward the negatively charged cathode. These positive particles are called **protons** (PROH-tahnz).

The charge of a proton is equal but opposite to the charge of an electron. A proton's charge is often described as 1+ (one plus). Although the charges of protons and electrons are equal but opposite, their masses are quite different. Protons have about 1840 times the mass of electrons.

 2. **Why did J.J. Thomson believe that electrons were present in the atoms of all elements?** _____

_____

In 1932, James Chadwick discovered still other particles in atoms. These particles are **neutrons** (NOO-trahnz). A neutron has no electric charge. But like the proton, the mass of a neutron is much greater than that of an electron. In fact, the masses of a proton and a neutron are almost the same.

The table below compares the three types of atomic particles to one another.

Particle Name	electron	proton	neutron
Particle Charge	1-	1+	0
Particle Mass	very small compared to the mass of a proton or a neutron	1840 times the mass of an electron	only very slightly greater than the mass of a proton

Check Your Understanding

3. Explain how electrons, protons, and neutrons are alike. How are they different? _____

_____

4. A particle has 4 protons and 3 electrons. What is its net charge? _____

5. A particle has 5 protons, 6 neutrons, and 5 electrons. What is its net charge? _____

6. A particle with 12 electrons has a -2 charge. How many protons does it have? _____

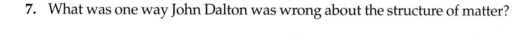

7. What was one way John Dalton was wrong about the structure of matter?

_____

8. How do the two types of charges interact with each other?

_____

9. A particle has 3 units of positive charge and 1 unit of negative charge. What is the overall charge of the particle? _____

10. What made J.J. Thomson believe that there were negatively charged particles passing through his glass tube?

_____

_____

What Do You Know?

12 Models Of The Atom

Key Words

nucleus: small region at the center of an atom where protons and neutrons are located

KEY IDEAS

The Dalton atom was a solid atom. The Thomson atom was mainly empty space containing positive and negative charges. In the Rutherford atom, the positive charges were in a nucleus. In the Bohr atom, electrons circle in orbits around the nucleus. In the charge-cloud model, electrons spin in all directions around the nucleus.

Early Greek scholars developed the idea that matter is made of particles that cannot be divided into anything smaller. They called the particles *atomos* after the Greek word meaning "not dividable." The idea of matter being made of small invisible particles was not widely accepted for centuries.

Dalton's Model of the Atom. In the early 1800s, John Dalton (1766-1844) described a model that could explain observations from several experiments. One such observation was that compounds have a constant composition. For example, in all samples of water, the mass of oxygen is always 8 times the mass of hydrogen. Dalton said that this law of constant composition could be explained by atoms. See Fig. 12-1.

Fig. 12-1

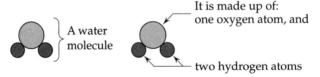

A water molecule

It is made up of:
one oxygen atom, and

two hydrogen atoms

SMALL SAMPLE OF WATER (1 molecule)

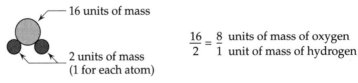

16 units of mass

2 units of mass
(1 for each atom)

$$\frac{16}{2} = \frac{8}{1} \frac{\text{units of mass of oxygen}}{\text{unit of mass of hydrogen}}$$

LARGER SAMPLE OF WATER (2 molecules)

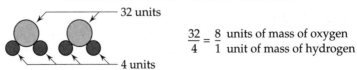

32 units

4 units

$$\frac{32}{4} = \frac{8}{1} \frac{\text{units of mass of oxygen}}{\text{unit of mass of hydrogen}}$$

The main ideas of Dalton's model were simple: All matter was made of atoms. Atoms were the simplest of all particles. They could not be divided into smaller particles. All atoms of the same element were alike. For example, they all had the same mass. Atoms of different elements were different; they had different masses. When a chemical reaction took place, the atoms in the reacting substances were rearranged.

 1. **See Figure 12-1 to answer this question. In a sample of water, there are 100 atoms of oxygen and 200 atoms of hydrogen. How will the mass of oxygen compare to the mass of hydrogen?**

_____

_____

Fig. 12-2

J.J. Thomson's Model. Recall that J.J. Thomson (1856-1940) discovered that atoms were not the smallest particles. Thomson proposed that an atom was made of positive charges from protons. Spread throughout these positive charges were the negative charges from electrons. See. Fig. 12-2.

Rutherford's Model. In 1909, Ernest Rutherford (1871-1937) did a famous experiment. Rutherford used alpha particles like tiny bullets. Each alpha particle carried two positive charges. Rutherford shot the alpha particles at high-speed at a thin sheet of gold. See Fig. 12-3.

Fig. 12-3

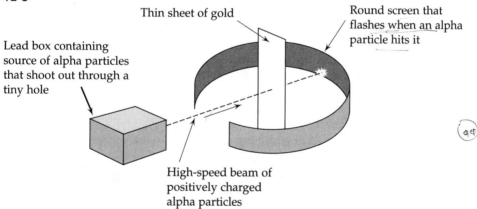

Thin sheet of gold

Round screen that flashes when an alpha particle hits it

Lead box containing source of alpha particles that shoot out through a tiny hole

High-speed beam of positively charged alpha particles

The Thomson atom was supposed to be mostly empty space. So Rutherford thought the alpha particles would pass right through the empty space in the atoms in the gold foil. Only if an alpha particle came close to a positive charge in a gold atom would the particle be bent slightly. Rutherford observed that what he expected to happen did happen in most cases.

However, something else happened that amazed him. The paths of about 1 out of 8000 alpha particles were bent in large angles (greater than 90°). See Fig. 12-4 on page 58. For this to happen, the alpha particles had to come close to a large positive charge. So Rutherford realized that the positive charges in an atom could not be spread throughout the atom.

Fig. 12-4

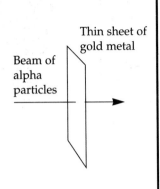

Beam of alpha particles

Thin sheet of gold metal

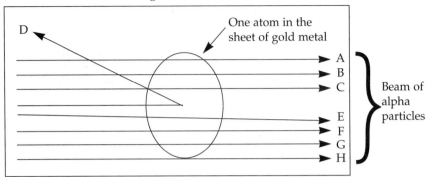

Thin sheet of gold metal

D

One atom in the sheet of gold metal

A
B
C

E
F
G
H

} Beam of alpha particles

Each of the letters from A to H shows the path of an alpha particle in the beam of alpha particles. Most of the positively charged alpha particles passed through the gold. The letter D shows the path of one alpha particle bent in a large angle. Rutherford thought that a large positive charge could repel the alpha particle and bend its path.

Fig. 12-5

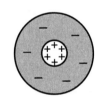

Rutherford decided that the positive charges of an atom had to be found in one tiny part of the atom. See Fig. 12-5. This part was called the **nucleus** (NOO-klee-uhs). Twenty-three years later, in 1932, a third particle, the neutron, was also found in the nucleus of the atom.

Bohr's Model. Niels Bohr (1885-1962) did important work that led to changes in Rutherford's atomic model. Bohr stated that electrons in atoms could exist only at certain energy levels. He believed electrons circled around the protons in the nucleus similar to the way planets travel around the sun. See Fig. 12-6.

Fig. 12-6

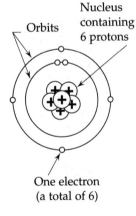

Orbits

Nucleus containing 6 protons

One electron (a total of 6)

✔ **2. How did Bohr's model differ from Rutherford's?** _____

_____

Bohr's model was partially successful. It described the locations of electrons in hydrogen atoms. However, the model could not explain the locations of electrons in atoms of other elements. You will read more about Bohr's model in Lesson 15.

Today's Model. Scientists working after Bohr made changes to his model. Today's model of the atom describes the locations of electrons in terms of probabilities. The probabilities are described by charge-cloud models. Figure 12-7 shows one example of a charge-cloud model. In this model, an electron is most likely to be found where the dots are close together. It is least likely to be found where the dots are far apart.

Fig. 12-7

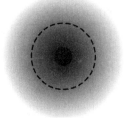

The most probable electron location is inside the dashed circle

Compare the five atomic models described in this lesson by studying the following table.

SUBATOMIC PARTICLE			PRESENCE IN ATOMIC MODEL				
Particle	Symbol	Charge	Dalton	Thomson	Rutherford	Bohr	Charge-Cloud
proton	p	+1	----	✓	✓	✓	✓
neutron	n	0	----	----	----	✓	✓
electron	e	-1	----	✓	✓	✓	✓

Check Your Understanding

Fill in the blanks in the following paragraph.

Two particles with equal but opposite charge are (3)_____ and (4)_____. The symbol for a neutron is (5)_____. The charge on a neutron is (6)_____. The Dalton atom contained (7)_____ subatomic particles. The Thomson and Rutherford atoms contained only (8)_____ and (9)_____. The atomic models that contain all three kinds of subatomic particles are the (10)_____ and the (11)_____.

What Do You Know?

12. What was Dalton's model of the atom successful in doing?

_____

13. Use carbon monoxide (a compound containing one carbon atom and one oxygen atom) to explain what the law of constant composition means.

_____

_____

14. What observations did not surprise Rutherford when he did his alpha particle experiment? What observations did surprise him?

_____

_____

15. In what way was Bohr's model of the atom similar to the model of the solar system? _____

_____

Lesson 13 Atomic Mass

Key Words

atomic number:	number of protons in each atom of an element
mass number:	number of protons and neutrons in an atom
isotope:	one of two or more kinds of atoms of the same element that differ from each other in their atomic masses
atomic mass unit:	a unit of mass equal to $1/12$ the mass of a carbon-12 atom

KEY IDEAS

While atomic number identifies the number of protons in an atom of an element, mass number gives the number of protons and neutrons. The atomic mass of the natural isotope of an element compares to the mass of the carbon-12 isotope. The carbon-12 isotope has been assigned a mass of exactly 12.

In 1913, H. G. J. Moseley used X-rays to find the actual charges of the nuclei of atoms. Moseley concluded that the positive charge on the nucleus of each succeeding element increased by one. He called this positive nuclear charge the atomic number.

Atomic Number. The **atomic number** of an element is the number of protons in each atom of the element. This number does not vary for a particular element. For example, all oxygen atoms have 8 protons. Atomic number has been given the short-hand notation Z. For oxygen, Z = 8.

Atomic numbers can be found in the table on page 84. Near the top of the left-hand column is a block containing Li. This is the symbol for lithium.

 1. **Look at the periodic table again. Where inside the block is the atomic number for lithium? What is the atomic number of lithium?**

Atoms are neutral. Thus, an atom must have the same number of electrons as protons. Therefore, the atomic number tells two things: (1) the number of protons in an atom and (2) the number of electrons in a neutral atom.

Neutrons and Isotopes. One or more neutrons are present in the nucleus of every atom except for most hydrogen atoms. In some elements, all the atoms have the same number of neutrons. All the atoms of fluorine, for example, have 10 neutrons.

There are other elements whose atoms differ in number of neutrons. Neon is an example. Its atoms are of two kinds. Some have 10 neutrons; others have 12 neutrons. Neutrons have mass. Therefore, an atom with 10 neutrons has less mass than an atom with 12 neutrons. The atoms with 10 neutrons are one of neon's isotopes. Those with 12 neutrons are another neon isotope. An **isotope** (EYE-suh-tohp) is one of two or more kinds of atoms of the same element having different atomic masses.

Mass number. If you add the number of protons to the number of neutrons in an atom, you get its **mass number**. The mass number for every fluorine atom is 19 (9 protons + 10 neutrons = 19). The capital letter A is used to identify mass number. For every fluorine atom, $A = 19$. Suppose you know both Z and A for a particular atom. By simply subtracting Z from A, you find the number of neutrons in the atom. In fluorine, for example, Z, the atomic number, is 9. If you subtract 9 from 19, you get 10, the number of neutrons in an atom of fluorine.

Chemists use a special notation to show both the mass number and the atomic number for an element. The mass number is written as a superscript in front of the symbol for the element. The atomic number is written as a subscript. See Fig. 13-1.

Fig. 13-1

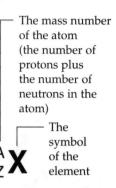

The mass number of the atom (the number of protons plus the number of neutrons in the atom)

The symbol of the element

$$_Z^A\text{X}$$

The atomic number of the element (the number of protons in the nucleus

 2. An atom of oxygen has 8 protons and 9 neutrons. Use this information to write the special notation for this oxygen atom. (The symbol for oxygen is O.) _____

There is another way of writing the mass number of an atom. You put the mass number after the name of the element. For example, neon-20 refers to the neon atom whose mass number is 20. (Its nucleus has 10 protons and 10 neutrons.)

Abundance of Isotopes. Consider a number of samples of neon taken from different places. All samples contain both neon-20 and neon-22. The number of neon-20 atoms compared to the number of neon-22 varies hardly at all. For every 100,000 atoms in any sample, there are about 90,920 neon-20 atoms. The rest of the atoms are neon-22. We say the abundance of neon-20 is 90.92 percent. (You get this number by dividing 90,920 by 100,000 and changing the decimal to a percent.)

Atomic Mass. The masses of atoms are very small. The following may give you an idea of just how small. A nickel has a mass of about 5 grams. To have a sample of carbon with a mass of 5 grams, it would take this number of carbon-12 atoms: 250,000,000,000,000,000,000,000.

Fortunately, chemists do not often have to deal with the mass in grams of a single atom. They use a special unit of mass to describe the masses of atoms. This unit is called the atomic mass unit (abbreviated μ).

The atomic mass unit is based on the mass of an atom of carbon-12. The mass of an atom of carbon-12 is assigned exactly 12 atomic mass units. The atomic mass of every other element can be given using this unit.

It is important to understand what the numbers mean in a table of atomic masses. Take, for example, the number 35.45 μ. This is the number listed as the atomic mass of chlorine. Samples of chlorine found in nature are not made of only one kind of chlorine atom. They are a mixture of two isotopes. One isotope is chlorine-35. The other is chlorine-37.

The mass of chlorine-35 is 34.97 μ. The mass of chlorine-37 is 36.97 μ. In every sample of chlorine, there are about 75 atoms of the first isotope for every 25 of the second. The lighter isotope is more abundant. The number found in the tables—35.45—is the average mass of the two isotopes. So, the atomic mass of an element is the average mass of its natural isotopes.

Look at the notation below for a certain isotope: $^{15}_{7}X$

The letter X is in the place of the symbol for a certain element. Note that the table below has two columns, I and II. Column I lists information revealed by the notation $^{15}_{7}X$. Column II tells how the information in Column I was determined. To complete Column I and Column II, a periodic table or table of atomic masses is used where necessary. See page 252.

	Column I	Column II
1	Number of protons is 7.	This is given by the subscript.
2	Number of neutrons is 8.	This is determined by subtracting the subscript from the superscript.
3	Z is 7.	Z is the atomic number. It is the same as the number of protons. (See line 1 above.)
4	A is 15.	A is the mass number. It is given by the superscript.
5	Atomic number is 7.	This is the subscript. (See line 1 above.)
6	Mass Number is 15.	This is given by the superscript. (See line 4 above.)
7	Symbol of element is N.	The periodic table shows that the element whose Z is 7 has the symbol N.
8	Atomic mass is 14.007.	This is given in the periodic table for the element whose atomic number is 7.
9	Name of the element is nitrogen.	This is from the table of atomic masses where the name of the element whose Z = 7 is given as nitrogen.

The paragraph refers to the atom whose notation is $^{11}_{5}X$. Use the information in the table above and a periodic table or a table of atomic masses to fill in the blanks for element $^{11}_{5}X$. See page 252.

The atomic number of the element is **(3)**_____ . The atom's mass number

is **(4)**_____ . There are **(5)**_____ neutrons and **(6)**_____ protons in

the atom. The symbol of the element is **(7)**_____ . The Z for the atom is

(8)_____ . The A for the atom is **(9)**_____ . The atom is an atom of the

element **(10)**_____ . The atomic mass of the element is **(11)**_____ . $^{12}_{5}X$

and $^{11}_{5}X$ are **(12)**_____ of the element.

What Do You Know?

13. In your own words, describe what the term *atomic number* means.

    _____

14. How do you describe an atom in which the number of protons is equal

    to the number of electrons? _____

    _____

15. One atom of a certain element has 9 neutrons. Another atom of the
    same element has 10 neutrons. **(a)** Besides the number of neutrons,

    what else will be different about these two atoms?_____

    _____

    (b) What will be the same? _____

    _____

16. An element has two isotopes. One is isotope A and the other is B. The
    amount of A is 65 percent. Tell in your own words what this means.

    _____

    _____

17. The atomic mass of an element is 36.00. How does the average mass of
    an atom of this element compare to the mass of a carbon-12 atom?

    Explain. _____

    _____

Lesson 14
The Nucleus

Key Words

alpha particle:	nucleus of a helium atom
beta particle:	electron
gamma rays:	very penetrating radiation like X-rays
nuclear reaction:	reaction in which a change takes place in the nuclei of atoms
half-life:	time it takes for half of the nuclei in a radioactive sample to disintegrate
transmutation:	process of one element changing into another during a nuclear reaction
transuranic element:	any element with an atomic number above 92

KEY IDEAS

When the nucleus of an atom is unstable, changes are likely to take place inside it. These changes include emitting, or giving off, an alpha particle, a beta particle, or gamma rays. Such changes make the nucleus more stable. When a nucleus emits an alpha particle or a beta particle, the original atom changes into an atom of a different element. Therefore, the atomic number, the mass number, and the atomic mass of the original element change.

Have you ever thought of becoming a medical assistant? If so, you may need to work with radioactive substances. Some radioactive substances are given directly to patients. Radioactive iodine, for example, is used both to study the thyroid gland and to treat thyroid problems.

The Stability of the Nucleus. Picture a wooden block on a table. Half of it is hanging over the edge. A change is likely to take place that will make the block more stable. If it falls off the table, it will land on the floor. In its new spot, it is more stable.

The nucleus of an atom can be unstable, too. Quite a few atoms have unstable nuclei. To become more stable, they tend to change. The reason a nucleus is unstable is that it lacks a balance between its protons and neutrons. Too many neutrons compared to the number of protons make a nucleus unstable. Too few neutrons compared to the number of protons also make a nucleus unstable. Atoms become more stable by emitting radiation.

Three kinds of radiation are commonly emitted by nuclei. They are alpha, beta, and gamma radiation. By emitting an alpha particle or a beta particle, the nucleus becomes more stable.

Alpha, Beta, and Gamma Radiation. Alpha particles and beta particles are charged particles. An **alpha particle** (AL-fuh) is the nucleus of a helium atom. The nucleus has two protons, so it has a charge of 2+. An alpha particle is written as $^4_2$He. The superscript shows that it is a helium nucleus with 2 protons plus 2 neutrons. The subscript shows the nuclear charge as 2+.

A **beta particle** (BAYT-uh) is simply an electron. It has a charge of 1-. Using superscripts and subscripts, an electron is written as $^0_{1-}$ e. Note that the mass number is written 0. This is because the mass of an electron is very, very small compared to the mass of a proton or neutron. Note also that the nuclear charge is written 1-. This charge comes from the electrons's 1- charge.

Gamma rays (GAM-uh) are a very penetrating form of radiation. They are more similar to X-rays than to alpha and beta radiation. Gamma rays have no charge.

Radioactive Elements. Elements whose nuclei emit alpha particles, beta particles, or gamma rays are radioactive. Radioactive elements give off energy when they emit these particles and rays. What actually happens when an alpha particle, a beta particle, or a gamma ray is emitted from the nucleus of an atom?

Alpha Emission. Emitting an alpha particle from a nucleus changes the original atom to a different atom. When an alpha particle is emitted, the nucleus of an atom changes. These changes are called **nuclear reactions** (NOO-klee-uhr ree-AK-shuhnz), and can be shown with a nuclear equation. To show how to write a nuclear equation, consider an atom of thorium-230. Its nucleus is unstable, and emits an alpha particle. Here are the steps in writing the nuclear equation for this reaction.

Step 1. *Write the symbol for the nucleus emitting an alpha particle. Then put an arrow to its right.* From the periodic table, find that the atomic number of thorium is 90, and that its symbol is Th. Therefore, thorium-230 can also be written $^{230}_{90}$ Th.

Equation 1 $\qquad ^{230}_{90}$ Th $\longrightarrow$

Step 2. *Write the symbol for an alpha particle after the arrow.*

Equation 2 $\qquad ^{230}_{90}$ Th $\longrightarrow$ $^4_2$ He

Step 3. *Write a plus sign after the alpha nucleus. Then write the letter X.* The X stands for the nucleus left after an alpha particle shoots out of a thorium-230 nucleus.

Equation 3 $\qquad ^{230}_{90}$ Th $\longrightarrow$ $^4_2$ He $+$ X

Step 4. *Determine the superscript for X.* Mass numbers are conserved. This means that the sum of all superscripts to the left of the arrow must equal the sum of all those to its right. We see that 230 minus 4 gives 226. Therefore, nucleus X can be written ^{226}X. As a check, note that $230 = 4 + 226$.

Step 5. *Determine the subscript for nucleus X.* The sum of all nuclear charges to the left of the arrow must equal the sum of all those to its right. Note that 90 minus 2 gives 88. Hence, nucleus X can now be written $^{226}_{88}X$.

Step 6. *Find what the symbol X stands for.* The atomic number of X is 88. From the periodic table, 88 is the atomic number of radon (Ra). Therefore, replace X in equation 3 with Ra.

$$\text{Equation 4} \qquad ^{230}_{90}\text{Th} \longrightarrow ^{4}_{2}\text{He} + ^{226}_{88}\text{Ra}$$

 1. **Review the 6 steps shown above. Then write an equation showing the emission of an alpha particle from polonium-218. Use the periodic table to help you.**

Beta Emission. Not all unstable nuclei emit alpha particles. Some emit a beta particle. Using superscripts and subscripts, the beta particle (an electron) is written $^{0}_{1\text{-}}\text{e}$. The emission of a beta particle from a lead-214 nucleus can be written as follows:

$$\text{Equation 5} \qquad ^{214}_{82}\text{Pb} \longrightarrow ^{0}_{1\text{-}}\text{e} + \text{X}$$

Equation 5 shows that lead-214 becomes nucleus X after it emits an electron. To complete Equation 5, you need a superscript for X and then its subscript. Finally you need its symbol. Following steps 4, 5, and 6 above, X must be $^{214}_{83}\text{Bi}$. (Remember, 82 minus -1 gives 83.) Replacing X in Equation 5 with $^{214}_{83}\text{Bi}$ gives

$$\text{Equation 6} \qquad ^{214}_{82}\text{Pb} \longrightarrow ^{0}_{1\text{-}}\text{e} + ^{214}_{83}\text{Bi}$$

Notice that the superscripts in Equation 6 balance: $214 = 0 + 214$. Note also that the nuclear charges balance: $82 = (1\text{-}) + 83$.

Gamma radiation often takes place when either an alpha particle or a beta particle is given off by a nucleus. However, gamma radiation causes no change in the atomic number or the mass number of the nucleus.

Half-life. When a radioactive nucleus emits a particle, the nucleus is said to decay. Radioactive nuclei do not decay all at once. No one can say when a particular nucleus will decay. However, the time needed for a large number of radioactive nuclei to decay can be measured. The time for half the nuclei in a radioactive sample to decay is called the element's **half-life**.

Plutonium-239 has a half-life of 24,000 years. A sample of plutonium today will be half gone after 24,000 years. A 16-gram sample today will have a mass of 8 grams in 24,000 years. After another 24,000 years, its mass will be only 4 grams, and so on.

Transmutation. Look back at Equation 4. Notice that thorium (Th) became radium (Ra). In other words, one element changed into another element. The process of one element changing into another element is called **transmutation** (trans-myoo-TAY-shuhn).

Transmutation never happens during chemical reactions. Only during nuclear reactions do changes take place in the nuclei of atoms.

2. **Are nuclear reactions involving alpha and beta emissions transmutation reactions? Why?** _____

_____

_____

_____

A very important natural transmutation takes place in uranium. Uranium is the heaviest natural element, atomic number 92. It can be changed into elements with atomic numbers 93 and 94. By beta emission $^{239}_{92}U$ is changed into neptunium (Np) and then into plutonium (Pu). These new elements are known as *transuranic* elements.

$$^{239}_{92}U \longrightarrow {}^{239}_{93}Np + {}^{0}_{1\text{-}}e$$

$$^{239}_{93}Np \longrightarrow {}^{239}_{94}Pu + {}^{0}_{1\text{-}}e$$

Transuranic elements (trans-yoo-RAN-ihk) are elements with atomic numbers above 92, the atomic number of uranium. Transuranic elements can be produced from various elements. They are not found in nature. Yet they are used in many different ways. For example, they are used in medicine and in smoke detection.

Energy Changes in Nuclear Reactions. When nuclear reactions take place, a tremendous amount of energy is released. This energy is what makes atom bombs and hydrogen bombs so destructive. The energy released when a chemical reaction takes place is very small by comparison.

Name	Type of Radiation	Charge	Symbols	Characteristics
Alpha	Particle	2+	α or ^4_2He	Nucleus of a helium atom
Beta	Particle	1-	β or $^0_{1-}e$	An electron
Gamma	Ray of Energy	none	Υ	Very penetrating ray like an X-ray

Check Your Understanding

3. Write a paragraph describing alpha, beta, and gamma radiation.

_____

_____

_____

Fill in the blanks to complete the paragraph.

The **(4)**_____ particle has a charge of 2+. The **(5)**_____ is an electron with a charge of **(6)**_____. The type of radiation that is energy rather than a particle is **(7)**_____. The symbol for an alpha particle is **(8)**_____ or **(9)**_____. The symbol $^0_{1-}e$ shows that a beta particle has a mass of **(10)**_____ and a charge of **(11)**_____. Υ is the symbol for **(12)**_____ radiation.

What Do You Know?

13. How does an unstable nucleus become more stable?

_____

14. Why do some nuclei found in nature give off alpha particles?

_____

15. What makes atomic nuclei unstable?

_____

16. What kinds of changes take place during a transmutation reaction?

_____

_____

17. Explain in words what the following equation means:

$$^{238}_{92}\text{U} \longrightarrow ^{4}_{2}\text{He} + ^{234}_{90}\text{Th}$$

_____

_____

_____

The symbol $^{r}_{s}\text{X}$ is used in a nuclear equation. The nucleus described by the symbol has 86 protons and 136 neutrons. Use the above information to answer questions 18 to 20. You may use a periodic table for help.

18. What does r stand for? _____

19. What does s stand for? _____

20. What is the symbol of the element that X stands for? _____

21. A radioactive element has a half-life of 64 years. You have a 48-gram sample of the element. What will be the mass of the element in the sample in 128 years? _____

Explain. _____

_____

22. Chemists who lived more than 200 years ago tried to change lead into gold. They tried to do this with various chemical reactions. They all failed. Explain why. _____

_____

_____

23. A certain element is radioactive. It emits alpha particles. Each of its nuclei contains 92 protons and 142 neutrons. Write a nuclear equation that describes the change that takes place by alpha emission.

_____

_____

24. Thallium-206 disintegrates by emitting beta particles. An incomplete equation for the change is the following: $^{206}_{81}\text{Tl} \rightarrow$? + ?. Complete the equation by replacing the question marks with the symbols for the missing particles.

_____

25. What is a transuranic element? _____

_____

_____

Lesson 15
Electron Energy Levels

Think of a book in a bookcase. A book can rest on a shelf, not between shelves. It takes energy to raise the book from a lower to a higher shelf because it takes energy to overcome the attraction of gravity. In somewhat the same way, an electron can be in a lower or higher energy level but not in between. It takes energy to move the electron to a higher energy level. This energy is given off when the electron returns to a lower level.

A Problem with the Rutherford Model. All moving electric charges lose energy. As an electron moves in an atom, the electron should lose energy. This energy loss would make the electron slow down. At slow speed, it could not resist the attraction between it and the positive nucleus. The electron would be pulled into the nucleus. The atom should collapse! Neils Bohr wondered why it didn't.

Bohr Energy Levels. The atom has a nucleus containing a proton. Around the nucleus, there are regions in space shaped like the surface of a ball. These regions are commonly called energy levels. They vary in size. The electron travels around the atom in one of these energy levels. In drawings, circles are used to show energy levels. See Fig. 15-1.

Energy levels are numbered by size (1, 2, 3, 4, and so on). The letter n stands for the number of an energy level. In the first energy level, n = 1. The lower the level, the smaller the energy of the electron in that level. See Fig. 15-1.

Fig. 15-1

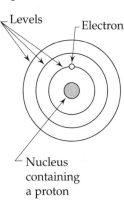

Levels
Electron
Nucleus containing a proton

An electron normally moves about the atom at only a certain energy level. It cannot have energy between two levels. An electron normally is found at the lowest energy level. When at that level, the atom is said to be in the **ground state** (grownd stayt).

An electron can jump from the first energy level to a higher level. To do this, the atom must absorb energy. The energy is needed to overcome the attraction between the negative electron and the positive nucleus. The atom has greater energy after the electron has moved to a higher energy level. When the electron is at a higher level, the atom is in an **excited state** (ehk-SEYET-ihd stayt).

Atoms do not stay in an excited state long. Their electrons soon return to the lowest energy level. Then the energy that had been absorbed is released from the atom. The energy is released as radiation, such as visible or ultraviolet light. The light from a neon sign comes from atoms of neon returning from an excited state to the ground state.

 1. **What happens when an atom in the excited state returns to the ground state?** _____

Bohr proposed an answer to the question of why atoms don't collapse. Atoms lose energy when electrons in excited atoms return to their lowest energy level. Once there, there is no lower energy level to which an electron can move.

Light Energy and the Bohr Model. Light is a form of energy. Fig. 15-2 shows a beam of white light separating into many colors. What appears on the screen is called a continuous spectrum. The word *continuous* means that one color blends into another. They blend like the colors in a rainbow.

Fig. 15-2

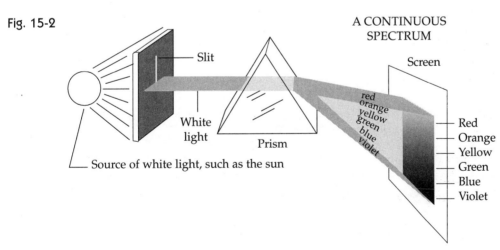

A CONTINUOUS SPECTRUM

Slit

Screen

White light

red
orange
yellow
green
blue
violet

Prism

Red
Orange
Yellow
Green
Blue
Violet

Source of white light, such as the sun

Fig. 15-3

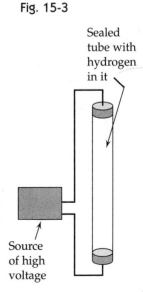

Sealed tube with hydrogen in it

Source of high voltage

Another kind of spectrum is called a line spectrum. It is made by putting a gas, such as hydrogen, into a sealed tube. See Fig. 15-3. At high voltage, hydrogen gives off a colored light.

Instead of a rainbow of colors, the spectrum now shows lines. See Fig. 15-4.

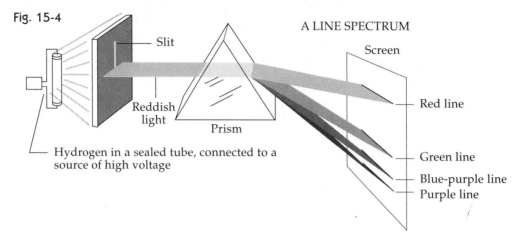

Fig. 15-4

A LINE SPECTRUM

Slit
Screen
Reddish light
Prism
Red line
Green line
Blue-purple line
Purple line
Hydrogen in a sealed tube, connected to a source of high voltage

Fig. 15-5

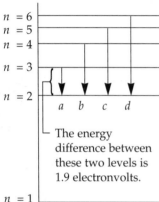

$n = 6$
$n = 5$
$n = 4$
$n = 3$
$n = 2$

a b c d

The energy difference between these two levels is 1.9 electronvolts.

$n = 1$

Each color (each wavelength) of light has an energy that is different from every other color. For example, purple light has more energy than red light. The energy of the red line in Fig. 15-4 is 1.9 electronvolts. (The electronvolt is a unit of energy.)

Bohr found that the 1.9 electronvolts of red light showed up somewhere else. It is the energy difference between the second and third energy levels. For the electron to jump from the second to the third level, the atom must absorb 1.9 electronvolts of energy. When the electron moves in the opposite direction—from the third to the second level—it gives off 1.9 electronvolts of energy. That is the energy of the red line in the spectrum of hydrogen. See Fig. 15-5.

TAKE ANOTHER LOOK

Figure 15-6 shows an electron and energy levels in an atom.

Fig. 15-6

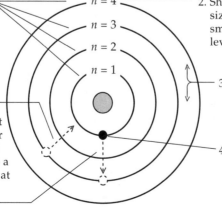

1. An atom has regions in space called "shells." A shell is like the surface of a hollow ball.

6. An atom will not remain excited. It will lose energy in the form of light when the electron returns to the lowest energy level.

5. If an atom absorbs the right amount of energy (from a source of heat or an electric spark), its electron will jump from a lower energy level to a higher level. When the electron is at the higher level, the atom is in an excited state.

$n = 4$
$n = 3$
$n = 2$
$n = 1$

2. Shells, or energy levels, are numbered by size. The level in which the electron has the smallest amount of energy is the first energy level ($n = 1$).

3. The electron can move about the atom only at certain energy levels. It cannot move at any other levels. It cannot move here.

4. The electron normally moves about the atom in the smallest shell, that is, at the lowest energy level, $n = 1$. When the electron is at the lowest level, the atom is in the ground state.

2. When an atom is in the ground state, what must happen for the atom to be in an excited state? What must happen for this atom to return to the ground state? _____

_____

_____

Complete the questions by filling in the blanks.

3. Electrons can occur only in specific _____.

4. An electron has the lowest energy in the energy level n = _____.

5. An electron has the greatest energy in the energy level n = _____.

6. An electron cannot occur _____ energy levels.

7. In a Bohr atom, how is one energy level different from another?

_____

8. What causes the light in a neon sign?

_____

9. How is a continuous spectrum different from a line spectrum?

_____

_____

10. Describe how a continuous spectrum and a line spectrum are produced.

_____

_____

11. Besides the way they look, how is purple light different from red light?

_____

12. What do the lines in a line spectrum tell about what is happening in an atom? _____

_____

13. In what two places does the energy 1.9 electronvolts show up?

_____

**What
Do You
Know?**

16 Electron Orbitals

KEY IDEAS

Electron orbitals play an important role in today's model of the atom. An orbital describes a region in space where an electron is most likely to be found. Different orbitals are found in every energy level. The average distance from the nucleus differs for each orbital.

In 1913, Bohr developed the model that explained the line spectrum of hydrogen. For this and other work in atomic structure, he received a Nobel Prize in physics.

Problems with the Bohr Model. Bohr's model worked well only for the atoms of hydrogen. Today's model has kept some of Bohr's ideas and thrown out others. In today's model, as in the Bohr model, electrons can have only certain energies. The letter n is still used. But Bohr's idea that electrons move around in circles like planets moving around the sun has been changed.

Exact Locations of Electrons Cannot be Known. Atoms are extremely small. Electrons are still smaller. Scientists began to realize that it is not possible to tell exactly where an electron can be found at any instant. Instead, the locations of electrons have to be stated in terms of probabilities.

To understand this idea, consider a honey bee collecting nectar from flowers. A bee must make trips to and from the hive. Suppose a camera is mounted above the hive. It is set to take one picture every minute over a five-hour period. In five hours, 300 pictures will be taken. See Fig. 16-1 on the next page. The dots in the figure show the position of the bee in each picture taken. Fig. 16-1 shows that the bee is most likely to be found near the hive.

Study Fig. 16-2. The probability of finding the bee within the dashed line in Fig. 16-2 is 90 percent. Thus, in 90 out of 100 times, the bee is likely to be within the dashed line. Fig. 16-2 shows a region in space where the bee is most likely to be found. It does not tell exactly where the bee can be found at any instant. This region in space may be compared to an electron orbital.

Fig. 16-1   The dots show the position of the bee.

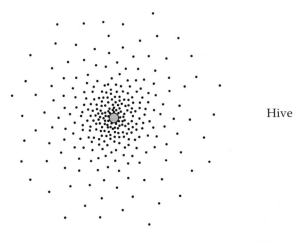

Hive

Fig. 16-2   The probability of finding the bee within the dashed line is 90%.

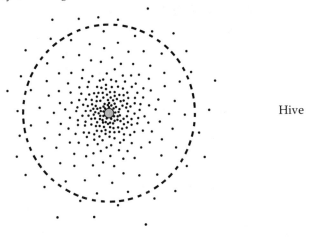

Hive

1. **Why is it more likely to find a bee near its hive?** _____

_____

Electron Orbitals.   An **electron orbital** (ee-LEHK-trahn AWR-biht-uhl), or orbital, describes a region in space where an electron is most likely to be found. There are four kinds of orbitals. Each kind has a different letter. The letters are *s*, *p*, *d*, and *f*. An orbital can hold no more than 2 electrons. Thus, it can be empty, have 1 electron, or have 2 electrons.

There is an *s* orbital at every energy level. An *s* orbital in the first energy level is called a 1*s* orbital. An s orbital in the second energy level is called a 2*s* orbital. When in a 2*s* orbital, an electron is farther from the nucleus on average than when it is in a 1*s* orbital.

What the Letter *n* Tells.   Look at the table on page 76. (1) The letter *n* tells the number of the energy level. It also tells the number of different *kinds* of orbitals at each energy level. (2) The *square* of the letter *n* tells the number of orbitals at any energy level. For example, for the third energy level, $n = 3$, and the square of $n = 9$. ($3^2 = 3 \times 3 = 9$).

Energy Levels and Electron Orbitals				
n	Number of kinds of orbitals	Letters for kinds of orbitals	Number of orbitals n^2	Number of orbitals and letter of each orbital
1	1	s	$1^2 = 1$	1: s
2	2	s p	$2^2 = 4$	4: s ppp (1 s an 3 p orbitals)
3	3	s p d	$3^2 = 9$	9: s ppp ddddd
4	4	s p d f	$4^2 = 16$	16: s ppp ddddd fffffff

Look at the last column. There is *one s* orbital in all four energy levels. In any energy level in which there are *p* orbitals, there are *three p* orbitals. (See the rows for n = 2, 3, and 4.) In any level in which there are *d* orbitals, there are *five d* orbitals. (See the rows for n = 3 and 4.) In any level in which there are *f* orbitals, there are *seven f* orbitals.

Recall that there are three possibilities for each orbital. It can be occupied by 0, 1, or 2 electrons. In the first energy level, there can be no more than 2 electrons. In the second energy level (with 4 orbitals), there can be no more than 8 electrons (2 x 4 = 8).

✓ 2. **What is the maximum number of electrons that can occupy the third energy level?** _____

   The fourth? _____

The table shows the number of orbitals of each kind and the maximum number of electrons in each kind of orbital in an energy level.

Letter of orbitals	Number of orbitals	Maximum number of electrons
s	1	2
p	3	6
d	5	10
f	7	14

Fill in the blanks in the sentences below.

3. Energy levels contain different numbers of _____.

4. Different _____ are used to represent each kind of orbital.

5. The number of *p* orbitals is _____.

6. There are seven _____ orbitals.

7. The maximum number of electrons in an *s* orbital is _____.

8. A maximum of 10 electrons is present in the _____ orbitals.

9. The maximum number of electrons in the *f* orbitals is _____.

10. In an energy level with 9 orbitals, the orbitals are lettered _____.

What Do You Know?

11. How is an electron orbit in Bohr's model different from an electron orbital in today's model? _____

_____

12. How many kinds of orbitals are in the third energy level? Explain.

_____

13. Suppose you know the number of *n*. What does this number tell you?

_____

14. How can you find out the number of orbitals at a particular energy level if you know the number of *n*?

_____

15. What is the maximum number of electrons possible at the fourth energy level? _____

Explain. _____

_____

16. How many *s* orbitals are there? _____ How many *p*? _____

How many *d*? _____ How many *f*? _____ What pattern do you see in these four numbers? _____

_____

Electron Distribution

KEY IDEAS

In atoms of different elements, the arrangement of electrons in orbitals follows a precise pattern. The electrons enter orbitals in order of increasing energy of the orbitals, beginning with the 1s orbital. At higher energy levels, some orbitals overlap.

Suppose you were preparing diagrams to show how electrons are arranged in elements. All goes well until you reach potassium, number 19. Something unexpected happens.

Electrons Are Arranged in Orbitals. In Fig. 17-1, the heavy dashed line divides the drawing into two parts. The left part shows numbers for n. The right has circles. Circles stand for orbitals. The circles have been color coded. For example, black circles are orbitals in the fourth energy level.

Fig. 17-1

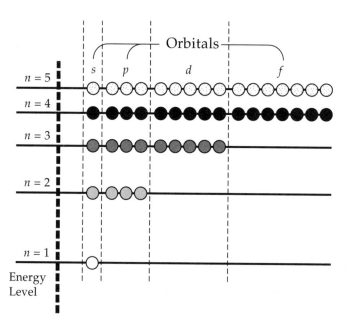

Recall that for $n = 2$, there are two kinds of orbitals (s and p). The total number of orbitals is 2^2, or 4 (one s and three p).

Fig. 17-2

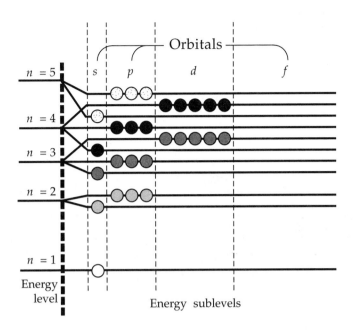

Fig. 17-2 shows more detail about the orbitals.

Look at $n = 2$. To the right of the dashed line, the horizontal line splits. The lower line goes to the $2s$ orbital, the higher line to the $2p$ orbitals. This is because the $2p$ orbitals have slightly higher energy than the $2s$ orbital.

The label at the bottom of Fig. 17-2 reads: "Energy sublevels." An energy sublevel is made of the set of orbitals (circles) on one line. The three $2p$ orbitals are at a slightly higher energy sublevel than the $2s$ orbital.

Fig. 17-3 shows how arrows are used to show electrons in orbitals. Circles with arrows in them are used to show how electrons are arranged in atoms.

 1. Draw an electron orbital to show the presence of

   one electron.

   two electrons.

Electron Configurations. To show how electrons are arranged in atoms, you could draw a diagram like Fig. 17-2. But to save space, draw an **electron configuration** (ee-LEHK-trahn kuhn-fihg-yoo-RAY-shuhn). An electron configuration shows in a single line the arrangement of electrons in an atom. To write an electron configuration, you must know the number of electrons in the atom. This is the same number as Z, the atomic number of the element. Fig. 17-4 shows electron configurations for the first 10 elements.

Fig. 17-3

Empty
orbital

Orbital with 1
electron in it

Orbital with 2
electrons in it

Fig. 17-4

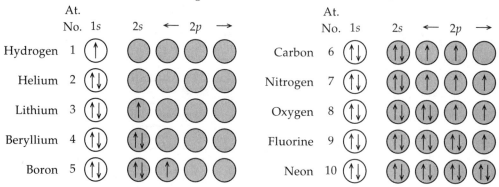

Electron Configurations For The First Ten Elements

Hydrogen (Z = 1). Because Z = 1, one electron needs to be placed. The 1s orbital has the lowest energy. Therefore, the electron goes into that orbital.

Helium (Z = 2). Two electrons must be placed. The 1s orbital has the lowest energy. Both electrons go into it.

Lithium (Z = 3). Three electrons must be placed. The 1s orbital is completely filled by the first two electrons. The third electron goes into the orbital of next higher energy. That is the 2s orbital.

Beryllium (Z = 4). The first two electrons occupy the 1s orbital. The next two, the 2s orbital.

Boron (Z = 5). After the 2s orbital, the next highest orbital is one of the 2p orbitals. The fifth electron goes into a 2p orbital. (Note that all three 2p orbitals have the same energy. Fig. 17-2 shows this by putting the orbitals on one line.)

Carbon (Z = 6). The first five electrons enter as for boron. The sixth electron is put into an empty 2p orbital. It is not put into the orbital that already contains an electron. All the p orbitals must be half filled before a second electron enters any of them. This is done because electrons repel each other.

Nitrogen (Z = 7). Here all three 2p orbitals are half filled.

Oxygen (Z = 8). The first seven electrons enter as they did for nitrogen. The eighth electron enters a half-filled 2p orbital. With two electrons in that orbital, it is completely filled.

The electron configurations of atoms for elements 9 to 18 can be drawn in the same way as those for elements 1 to 8.

Consider now elements of higher atomic number.

Fig. 17-5

Electron Configurations For Elements 19, 20, 21

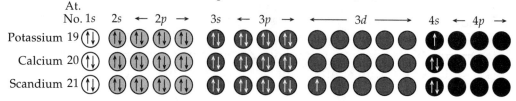

Potassium (*Z* = 19). See Fig. 17-5. Note that 18 of the 19 electrons go into the *s* and *p* orbitals of energy levels 1, 2, and 3. The nineteenth electron does the unexpected. It skips all five 3*d* orbitals. It enters the 4*s* orbital.

You can see the reason for this in Fig. 17-2. Observe that the 4*s* orbital has lower energy than the 3*d* orbitals. Because the nineteenth electron goes into the 4*s* orbital, the atom has lower energy than if the electron had gone into a 3*d* orbital. This happens because of the way electrons interact with each other in atoms with many electrons.

Calcium (*Z* = 20). The twentieth electron also enters the 4*s* orbital.

Scandium (*Z* = 21). Note from Fig. 17-2 that after the 4*s* orbital, the orbitals having the next highest energy are the 3*d* orbitals. Therefore, the twenty-first electron goes into a 3*d* orbital. Fig. 17-5 shows this configuration.

Fig. 17-6 shows electron configurations for elements 14, 18, and 23.

Fig. 17-6

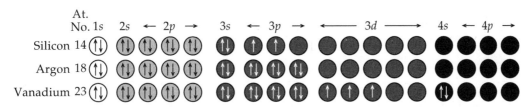

Check Your Understanding

2. Vanadium has circles with 2, 1, and 0 arrows. Explain what this means.

   _____

   _____

3. Show the electron configuration for sulfur, atomic number 16.

   _____

4. For *n* = 3, how many different kinds of orbitals are there? _____

5. For *n* = 4, what is the total number of orbitals? _____

6. Show the electron configuration for nickel, z = 28.

   _____

What Do You Know?

18 The Periodic Table

Key Words

period: horizontal row of elements in a periodic table

group: vertical column of elements in a periodic table

KEY IDEAS

In the periodic table, the elements are arranged according to increasing atomic number. In the table, the elements are also arranged into periods and groups. The elements in a group have similar properties.

Many early chemists tried to arrange elements in groups according to their properties. In 1869, Dimitri Mendeleev of Russia arranged the 73 known elements into a table according to increasing atomic masses. This table is the basis of the modern periodic table.

The Periodic Table Today. Fig. 18-1 shows a line of 18 elements. The line starts with Li, whose atomic number, Z, is 3. It ends with Ca, whose atomic number is 20. All 18 elements are in increasing order of Z. There is a pattern to this arrangement. Every eighth element has similar properties. The properties repeat after every eighth element. Elements with arrows of the same length above them have similar properties.

Fig. 18-1

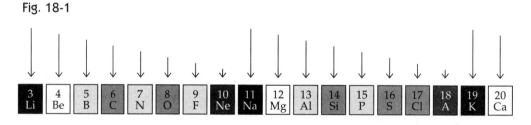

Fig. 18-2 shows the same elements arranged differently. The second set of 8 elements, starting with Na, has been put under the first set. This puts elements with the same properties in a column. Elements with similar properties are stacked above and below one another.

Fig. 18-2

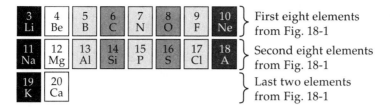

First eight elements from Fig. 18-1

Second eight elements from Fig. 18-1

Last two elements from Fig. 18-1

The last element in Fig. 18-2 is element 20. You would expect element 21 to go under element 13. However, its properties do not match those of element 13. In fact, the properties of none of the 10 elements from numbers 21 through 30 match those of any of the elements in Fig. 18-2. Therefore, none of those 10 elements can be placed under the elements in Fig. 18-2. See Fig. 18-3.

Fig. 18-3

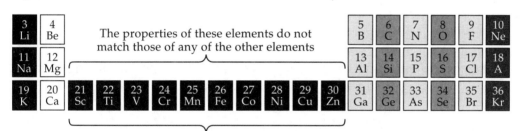

Called "transition elements." They form a kind of bridge between the elements to either side.

Element 31 has properties that match those of element 13. That is why it appears under element 13 in Fig. 18-3. The properties of the next 5 elements match the properties of elements 14 through 18. They are therefore put beneath elements 14 through 18. The 10 elements between elements 20 and 31 are known as transition elements.

See Fig. 18-4 or page 252 for the complete periodic table. A row of elements in a periodic table is called a **period** (PIHR-ee-uhd). There are seven periods. The number of each period is shown in a black circle. A column of elements in a periodic table is called a **group** (groop). The number of each group is shown above each column.

The elements between groups IIA and IIIA are the transition elements. They are not given individual group numbers here. You will not study them in detail.

Fig. 18-4

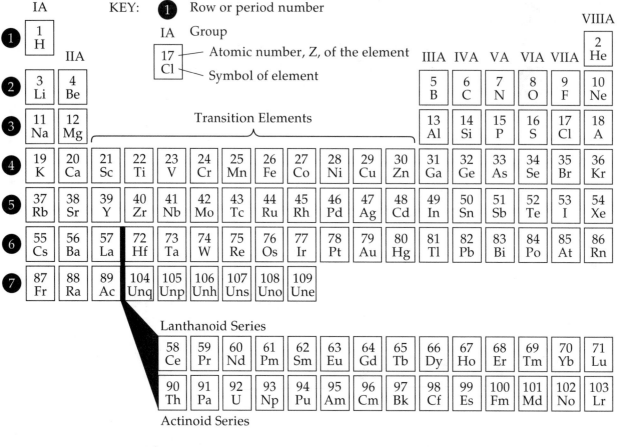

 1. List the symbols of all elements in the second period.

_____

 2. List the symbols of all elements in group IIA.

_____

Hydrogen, a Special Element. Figs. 18-1, 18-2, and 18-3 did not include the two lightest elements. They are hydrogen (H) and helium (He). In Fig. 18-4, hydrogen appears under IA. The 6 elements below hydrogen are extremely reactive metals. They are called alkali metals. Hydrogen is not an alkali metal. This is why there is space separating hydrogen from the alkali metals.

Helium. Helium, the next lightest element, is at the top of Group VIIIA. It and the 6 elements below it are all members of the same family. They are called the noble gases because until recently it was believed that no other elements reacted with them.

The Fourth and Fifth Periods. The fourth period has 18 elements, including 10 transition elements. Chemists chose this name because these fairly active metals form a series, or transition, between the active metals and the nonmetals. The fifth period, like the fourth period, has 18 elements, including 10 transition elements.

The Sixth and Seventh Periods. The sixth period starts out like the fifth period. However, beginning with element 58, there are 14 elements that are quite similar. These elements have similar metallic properties, are fairly rare, and are often found together. Their electron configurations are fairly complex. These 14 elements are called the lanthanoid series. The lanthanoids do not appear in the main body of the table. They appear below it.

The seventh period is like the sixth in one way. Beginning with element 90, there is another group of 14 elements. These elements, like the lanthanoids, have similar properties. This second set is called the actinoid series. The actinoids appear below the lanthanoids. The lanthanoids and actinoids are called the inner transition elements.

 3. How are the lanthanoid and actinoid series alike? Different?

_____

_____

If you look at a periodic table you will see that each square contains the atomic number, and the symbol of an element.

Metals, Nonmetals, and Metalloids. Most of the elements are metals. A few are nonmetals. Even fewer are metalloids. See Fig. 18-5 at the top of the next page. A heavy, zigzag line separates the metals and the nonmetals. The few metalloids fall along this line.

Metals are good conductors of heat and electricity. Many of them have luster, or shine. They are also malleable, or able to change shape without breaking.

Nonmetals are gases, liquids, or crystalline solids. They lack the properties of metals.

Metalloids are sometimes called semimetals. They have properties that fall between those of metals and nonmetals.

Trends Within a Group. See Fig. 18-4. Consider how elements are different within a column, or group. Metals become more metallic going from top to bottom. For example francium (element 87) is a more reactive metal than the element above it, cesium. Potassium (element 19) is more reactive than sodium (element 11).

Consider the trend within a nonmetallic family. In the halogen family (group VIIA), elements become more nonmetallic from bottom to top. Fluorine (element 9), for example, is a more reactive nonmetal than the elements below it. Chlorine (element 17) is a more reactive nonmetal than the elements below it.

Fig. 18-5

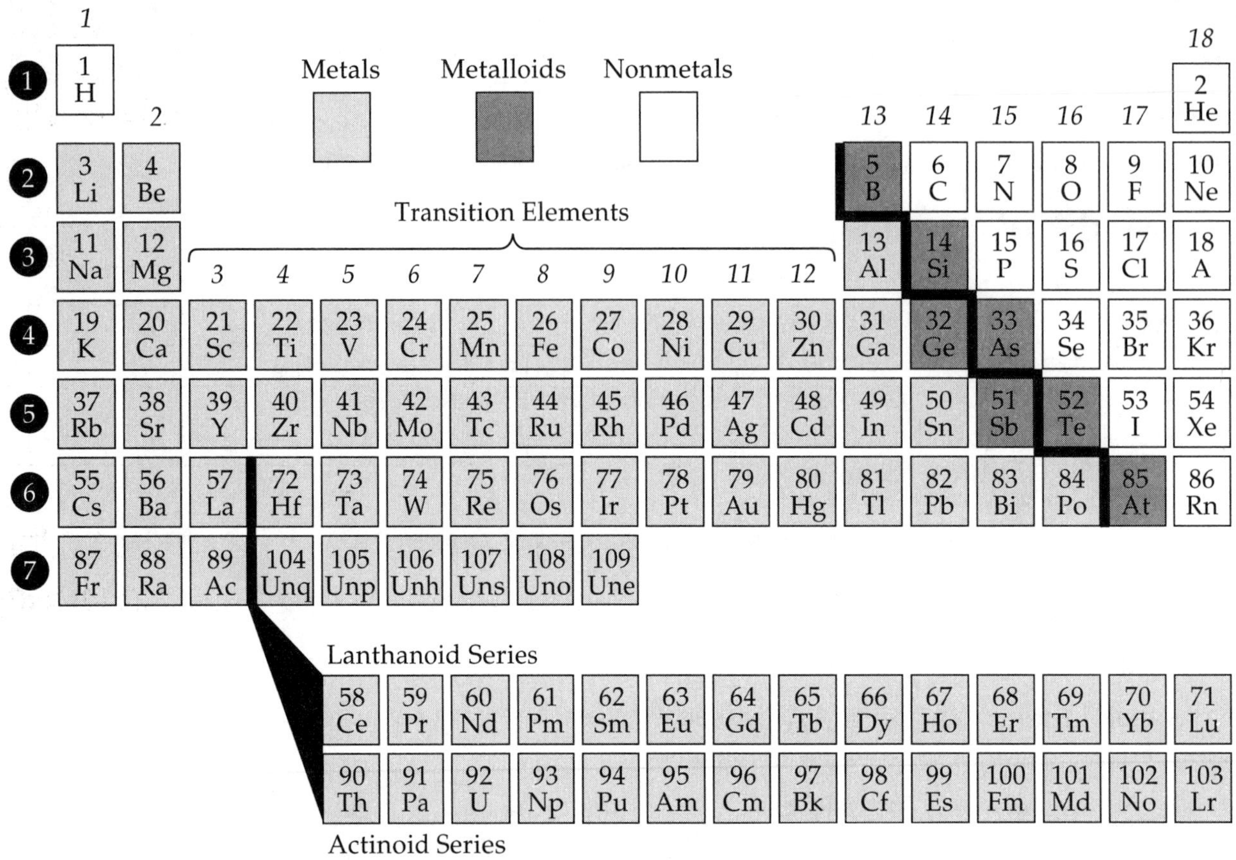

Periodic Trends. See Figs. 18-4 and 18-5. Consider how elements differ in their properties across a period. The first element in a period is in group IA. Group IA metals are alkali metals. These elements are strongly metallic. Going from left to right across a period, elements become less strongly metallic. The elements in group VIIA, the halogens, are the strongest nonmetals. In period 2, fluorine (element 9) is more nonmetallic than oxygen (element 8). Oxygen is more nonmetallic than nitrogen (element 7).

The Periodic Table and Electron Configurations. The properties of elements are the result of the way electrons are arranged in the outer orbitals of atoms. Elements of atoms whose electrons are arranged in the same way in their outer orbitals have similar properties. The alkali metals have similar properties because their outer orbitals contain one electron. The halogens, group VIIA, are similar because each has seven outer-orbital electrons.

Some textbooks now label the eight A groups and the ten groups of transition elements by numbers 1 through 18. These numbers are in *italics* at the top of each of the eighteen groups in Fig. 18-5.

In the periodic table in Fig. 18-6, the lanthanoids and actinoids have been placed in the main body of the table. This makes the table wide and the boxes small. Only a few elements and their atomic numbers are shown.

Fig. 18-6

	IA																		VIIIA
1	1 H	IIA												IIIA	IVA	VA	VIA	VIIA	2 He
2	3 Li														5 B				
3		12 Mg																	18 A
4		20 Ca														31 Ga				
5																			54 Xe
6		57 La	←		Lanthanoid Series			→	72 Hf										
7			←		Actinoid Series		→												

Use the periodic table, Fig. 18-4, to answer the questions below.

4. Elements 2, 18, and 54 are in group _____.

5. An element in period 5 is _____.

6. Two elements in period 6 are _____ and _____.

7. Element 20 is in group _____ and period _____.

8. Period 7 contains the _____ series.

What Do You Know?

9. In Fig. 18-2, there's a difference of eight between an element and the element above it. Why doesn't element 21 go under element 13?

_____

10. Why is there extra space between hydrogen (element 1) and the element under it in Fig. 18-4? _____

11. What family begins with helium (element 2)? _____

12. Which would you expect to be more reactive, potassium (element 19) or sodium (element 11)? _____

13. Which would you expect to be more reactive, chlorine (element 17) or bromine (element 35)? _____

14. In terms of electron configurations, how are the alkali metals similar to one another? _____

Summary

- Ideas about the structure of the atom have changed during the past 200 years.

- Dalton believed matter consisted of small, indivisible particles. J.J. Thomson discovered that atoms contained electrons.

- Rutherford found that the positive charge in an atom could be found in a tiny region at the center of the atom.

- Bohr realized that electrons could occupy only certain energy levels but not others.

- Today's model of the atom includes the idea that the position of an electron cannot be known exactly at any particular time.

- All atoms of the same element have the same number of protons. However, not all atoms of the same element have the same mass.

- Atoms of the same element may differ from one another in the number of neutrons in the nucleus. Atoms of the same element having different numbers of neutrons are called isotopes.

- The masses of atoms are measured in atomic mass units (amu). One atomic mass unit is equal to $1/12$ the mass of a carbon-12 isotope.

- The nuclei of some atoms are unstable because they have too many or too few neutrons compared to the number of protons. By emitting particles, the nuclei can become more stable.

- Alpha and beta particles are often emitted from unstable nuclei. The nucleus of the atom from which an alpha or a beta particle is emitted changes to the nucleus of another element.

- The kind of change in which a particle is emitted from an atom's nucleus is called transmutation. A nuclear equation is used to represent changes in the nucleus of an atom.

- Electron orbitals are regions in the space within an atom where the probability of finding an electron can be determined. The way electrons are arranged in orbitals is called the atom's electron configuration.

- The periodic table organizes the elements according to similarity in properties. Elements with similar properties appear one above the other in groups.

For Your Portfolio

1. In addition to water, there is another compound of hydrogen and oxygen called hydrogen peroxide. A molecule of hydrogen peroxide has two oxygen atoms and two hydrogen atoms. See Fig. 3-A. Prepare a drawing like Fig. 12-1 on page 56 to show that in different samples of hydrogen peroxide, the mass of oxygen will be 16 times the mass of hydrogen.

    Fig. 3-A

    A hydrogen peroxide molecule

    16 units of mass for each oxygen atom

    1 unit of mass for each hydrogen atom

2. Demonstrate models of the atom using a group of eight or ten students. Half the students are electrons. The other half are protons. **(a)** Take positions in the classroom to show how electrons and protons were arranged in a J.J. Thomson atom. **(b)** In the same way, model the Rutherford atom and the Bohr atom.

3. The class should form groups of two. Have one student describe to the other the arrangement of orbitals in energy sublevels. A drawing may also be used as part of the description.

4. Choose four students, two of them to act as protons and two as neutrons. Demonstrate for the rest of the class the difference between the nuclei of helium-3 and helium-4.

5. Make two drawings, one showing an atom of hydrogen in the ground state and the other showing it in an excited state. Show on the drawing what happens to the atom when it absorbs energy. Also show what happens when the atom loses energy.

6. Form pairs. **(a)** Have one student explain what happens to the atomic number of a nucleus when the nucleus gives off an alpha particle. **(b)** Have the other student explain what happens to the atomic number of a nucleus when the nucleus gives off a beta particle. Paper and pencil may be used as aids.

7. Draw an orbital diagram for an atom that has five electrons. Make a second orbital diagram showing titanium, element 22.

8. Complete the equation for the emission of a beta particle by lead -210.

$$^{210}_{82}\text{Pb} \longrightarrow ? + ?$$

Fill in the blank with the letter of the term that best completes the statement.

_____ 1. Two atoms are isotopes if they are atoms of

   a. the same element with a different number of neutrons.
   b. different elements with a different number of neutrons.
   c. the same element with the same number of protons.
   d. different elements with the same number of neutrons.

_____ 2. When its electrons are in the lowest possible energy levels, an atom is in the

   a. confused state.   c. excited state.
   b. ground state.   d. transmuted state.

_____ 3. In an atom, two particles that have

   a. opposite charges are protons and neutrons.
   b. opposite charges are electrons and neutrons.
   c. like charges are protons and electrons.
   d. nearly equal masses are protons and neutrons.

Column I contains definitions of some terms. Column II contains the terms being defined. On the line at the left of each definition in Column I, write the letter of the term being defined in Column II.

Column I

_____ 4. the number of protons in the nucleus of an atom

_____ 5. the number of protons in the nucleus of an atom plus the number of neutrons

_____ 6. the average mass of the naturally occurring isotopes of an element based on the mass of a carbon-12 isotope whose mass is set at exactly 12

_____ 7. the part of an atom where the protons and neutrons are found

_____ 8. the region in an atom where an electron is found, according to today's model of the atom

_____ 9. the nucleus of a helium atom

Column II

   a. orbital
   b. mass number
   c. atomic number
   d. nucleus
   e. atomic mass
   f. alpha particle

Write a paragraph to answer one of the following questions.

10. a. What is the difference between a continuous spectrum and a line spectrum?

    b. Explain how the periodic table is arranged. In doing so, use the following terms: _period, group, atomic number, atomic mass, metal, nonmetal, metalloid, transition elements, lanthanoids, actinoids._

Chemical Bonds

Have you ever used super-strong glue to attach things such as the parts of a model? If you accidentally get some glue on your fingers, you could have a sticky problem! It's hard to separate things that are stuck together with super-strong glue.

The atoms that make up all matter can join together to form compounds. They're joined by forces that are stronger than the strongest glue.

Groups of atoms are held together by electrical forces. These forces result from the attraction between atomic nuclei and electrons. Strong bonds form between atoms when they attract the same electrons at the same time. These bonds contain a great deal of energy and are hard to break. They also determine the kinds of compounds formed, the properties of these compounds, and even their shapes.

Ionic Bonds

chemical bond:	force that holds two atoms together
chemical formula:	group of symbols used to show the kinds and numbers of atoms in a compound
ionic bond:	chemical bond formed when electrons are transferred from the outer energy level of one atom to the outer energy level of another atom
ion:	atom that develops an electric charge when it loses or gains electrons to fill its outer energy level
cation:	positive ion formed when a neutral atom loses electrons to have a filled outer energy level
anion:	negative ion formed when a neutral atom gains electrons to have a filled outer energy level
crystal:	solid in which particles are in a regular, repeating pattern

KEY IDEAS

Two atoms can be held together in a compound when electrons from the outer energy level of one atom are shifted to the outer energy level of the other. This movement of electrons forms an ionic bond. Compounds with ionic bonds have certain important properties.

Photosensitive glass darkens in the sun, reducing the amount of sunlight that passes through. Chemists make photosensitive glass by mixing silver chloride crystals with the glass. The silver and chlorine atoms separate from each other, forming charged particles that react with the sun's rays to darken the glass.

Forming Compounds. Most atoms do not exist by themselves. Instead, they join with other atoms to form compounds. The way the atoms are joined together determines the properties of a compound.

When atoms combine, the positions of their electrons change. This results in a force called a **chemical bond** (KEHM-ih-kuhl) that holds the atoms together. The number to remember for chemical bonds is 8. Most atoms join in chemical bonds in order to have 8 electrons in their outer energy level. An outer electron energy level with 8 electrons is said to be filled.

Sometimes atoms must either lose or gain electrons to have a filled outer energy level. For example, the sodium (Na) and chlorine (Cl) atoms that make up table salt (NaCl) are joined by a chemical bond.

NaCl is a chemical formula. A **chemical formula** (KEM-ih-kuhl FAWR-myoo-luh) is a group of symbols showing the kinds of atoms and the numbers of atoms of each kind in a compound. Look at Fig. 19-1. The Na atom has 11 electrons, but only 1 in its outer energy level. The easiest way for Na to have a filled outer energy level is to lose that 1 electron.

Fig. 19-1

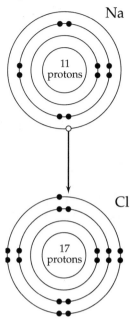

Chlorine (Cl), however, is just the opposite of sodium. Count the 7 electrons in its outer energy level. To have a filled outer energy level of 8 electrons, Cl needs 1 more electron.

Ionic Compounds. Na must lose an electron and Cl must gain one for the two atoms to join together. This is shown by the arrow in Fig. 19-1. The Na and Cl form NaCl. Now Na has a filled outer energy level of 8 electrons, and so does Cl. When electrons shift from the outer energy level of one atom to the outer energy level of another atom, an ionic bond forms. An **ionic bond** (eye-AHN-ihk bahnd) is a chemical bond formed when electrons move from the outer energy level of one atom to the outer energy level of another atom. An ionic bond holds together the atoms in an ionic compound.

 1. **How many electrons are needed to completely fill the outer energy level of most atoms?** _____

Atoms change when they enter into an ionic bond. This change is what gives an ionic compound its properties. When atoms lose or gain an electron, their number of protons and electrons is no longer equal. Such atoms now have an electric charge. Any atom that develops an electric charge when it loses or gains electrons to have a filled outer energy level is called an **ion** (EYE-uhn).

Ions can have either a positive charge or a negative charge. When a Na atom loses 1 electron, 11 protons and 10 electrons are left. The Na ion formed has a positive charge of 1 (Na^{1+}). Any atom that has a positive charge is known as a **cation** (KAT-eye-uhn).

When a Cl atom forms an ionic bond, it gains 1 extra electron. With 18 electrons and only 17 protons, the Cl ion has a negative charge of 1 (Cl^{1-}). A negative ion formed when an atom gains one or more electrons is called an **anion** (AN-eye-uhn).

Because the ions in an ionic compound have opposite charges, they attract one another. This attraction is the force of the ionic bond.

 2. **Explain the difference between a cation and an anion.**

_____

_____

Properties of Ionic Compounds. Many substances with ionic bonds, such as NaCl (table salt), have a crystal structure. A **crystal** (krihs-tuhl) is a solid in which the particles are arranged in a regular, repeating pattern. The strong ionic bonds hold the cations and anions firmly in place. But ionic compounds usually have high melting points and dissolve in water. When a crystal dissolves in water, the crystal breaks up and the ions move about freely. A mixture of ions in water can conduct an electric current. This property of ions is used in batteries. In batteries, an electric current is passed from one pole to the other, carried by the ions.

Fig. 19-2 summarizes the relationships among chemical bonds.

Fig. 19-2

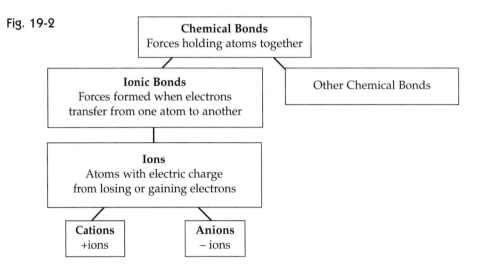

Check Your Understanding

Write a sentence explaining the connection between each pair of terms.

3. chemical bond, force _____

4. ionic bond, chemical bond _____

5. cation, positive _____

6. anion, negative _____

7. Complete the electron structure of the Na and Cl atoms in Fig. 19-3. Use an arrow to show how an ionic bond would form between them.

Fig. 19-3

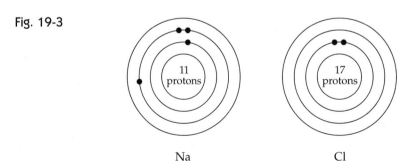

8. Which structures of an atom take part in a chemical bond?

   _____

9. How many electrons does it take to fill the outer energy level of most atoms? _____

10. How does the electron structure of atoms change when they form chemical bonds? _____

    _____

11. What is the difference between a sodium atom (Na) and a sodium ion (Na^{1+})? _____

    _____

12. An atom of potassium (K) has 19 protons: 18 electrons in its inner electron energy levels and 1 electron in its outer energy level. What is the easiest way for this atom to get a filled outer energy level?

    _____

    What is formed? _____

13. Is Li^{1+} a cation or an anion? _____
    Why? _____

14. Name two properties of many substances containing ionic bonds.

    _____

15. Where in the periodic table (see Lesson 18) are the elements that tend to form positive ions? Negative ions? Explain.

    _____

    _____

    _____

    _____

Covalent Bonds

KEY IDEAS

Two atoms can be held together in a compound when they share two electrons in their outer energy levels. This type of joining forms a covalent bond. Compounds with covalent bonds have certain important properties.

More compounds of carbon are known than compounds of all other elements combined. Since carbon forms compounds by covalent bonds—whether in gasoline, nylon, or protein—you see one reason why covalent bonds are so important.

Covalent Compounds.   Some compounds form ionic bonds. But most compounds are held together by a different kind of chemical bond. Indeed, any product made from a plant or an animal—from a cotton shirt to a tuna sandwich—is made of many carbon atoms. These atoms are joined together by a chemical bond in which the atoms share electrons. A chemical bond formed by the sharing of a pair of electrons between two atoms is called a **covalent bond** (KOH-vay-luhnt). Like ionic bonds, covalent bonds are formed by the attraction between electrons and the nuclei of atoms. But in covalent bonds, the atoms fill the outer electron energy level by sharing electrons, not by transferring them. No ions are produced.

Fig. 20-1

✓ **1.   How many electrons are shared in a covalent bond?** _____

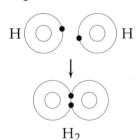

The substance scientists call the fuel of the future for cars actually has the simplest covalent bond. This is hydrogen gas (H_2), which burns more cleanly than other fuels. The hydrogen atom must join with another atom to fill its outer electron energy level. See Fig. 20-1. The filled outer energy level of hydrogen can hold just 2 electrons. By sharing 2 electrons in a covalent bond, each H atom has a filled outer energy level. A group of two or more atoms joined together by covalent bonds is known as a **molecule** (MAHL-ih-kyool). Ionic compounds do not form molecules.

The pair of electrons in a hydrogen molecule is shared equally by each hydrogen atom. Unlike ions, the atoms in this molecule have no electric charge. In fact, substances with covalent bonds are usually poor conductors of electricity.

Molecules are neutral in charge. Sometimes, however, the atoms forming the covalent bond can have a small electric charge. Having a small electric charge usually happens when atoms that easily lose electrons combine with atoms that easily gain electrons. Look at the molecule of hydrogen chloride (HCl) in Fig. 20-2. The H and the Cl atoms each get a filled outer energy level by sharing the electron pair. But the Cl atom pulls the electrons much more strongly than the H atom. So the electron pair is actually closer to the Cl atom in this molecule.

Fig. 20-2

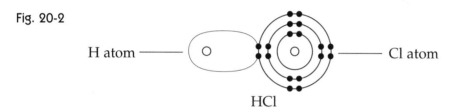

H atom ——— ⬭ ——— Cl atom

HCl

A **polar** (POH-luhr) covalent bond is a covalent bond in which the shared electron pair is pulled closer to one atom. This means that the electrons are not equally shared. The atom that pulls more strongly on the electrons has some negative charge. The atom that pulls less strongly on the electrons has some positive charge. This type of molecule is called a polar molecule. HCl is one example. The Cl atom has some negative charge, while the H atom has some positive charge.

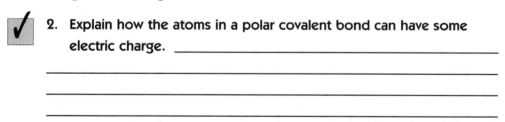

2. Explain how the atoms in a polar covalent bond can have some electric charge. _____

_____

_____

_____

The polar covalent bonds in water (H_2O) molecules explain why many other substances dissolve in water. The O atom strongly pulls the electron pair it shares with each H atom closer. The O atom has some negative charge. If, for example, salt (NaCl) is placed in H_2O, the positive Na ions are pulled toward the negative O atoms of the H_2O molecules. The negative Cl ions go toward the positive H atoms of the water molecules. This breaks down the NaCl crystal shape and the salt easily dissolves in the water. HCl molecules also dissolve easily in water, making hydrochloric acid.

Fig. 20-3 summarizes the relationships among bonds in covalent compounds.

Fig. 20-3

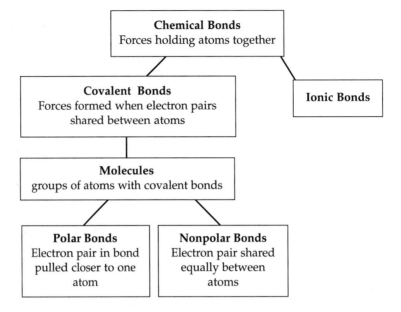

Check Your Understanding

Complete each of the following sentences by circling the correct underlined term.

3. In ionic / covalent bonds, a pair of electrons / protons is shared by two atoms.

4. The outer energy levels of the tiny atoms hydrogen and helium (atomic number 2) are filled by 2 / 8 electrons.

5. No ions are present in molecules / crystals.

6. In a polar covalent bond, the electron pair is shared equally / unequally between the two atoms.

7. Fig. 20-4 shows the electron structure of fluorine (F). In the space provided, show how fluorine joins with hydrogen to form a molecule of hydrogen fluoride (HF).

Fig. 20-4

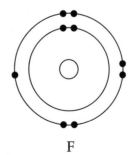

F

8. What structures of an atom take part in a covalent bond?

   _____

9. How is a covalent bond formed? _____

   _____

10. How many covalent bonds are present in the
    molecule shown in Fig. 20-5? _____
    What is the total number of electrons in these
    covalent bonds? _____

Fig. 20-5

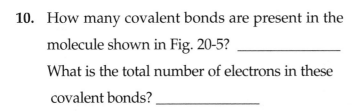

11. Would it be correct to say a "molecule of NaCl"? _____
    Explain your answer. _____

    _____

    _____

12. What kind of chemical bonds are present in substances made of carbon
    atoms linked together? _____

13. What is the difference between a covalent bond that is polar and one
    that is not? _____

    _____

    _____

14. Is the covalent bond between the chlorine (Cl) atoms in a molecule of
    chlorine gas (Cl_2) polar? Explain. _____

    _____

    _____

    _____

15. Why do polar substances dissolve well in other polar substances?

    _____

    _____

    _____

    _____

Lesson 21
Molecular Shapes

Key Words

bond angle: angle between two covalent bonds

KEY IDEAS

Molecules have different shapes depending on the kinds of atoms bonded together and the electron pairs present. Molecular shapes affect the properties of molecules.

The properties of methane gas make it a very useful fuel. Methane is colorless, is almost odorless, and combines with oxygen. In fact, the natural gas used in homes for cooking and heating is methane. Scientists believe that huge amounts of methane gas may be trapped in ice under the ocean floor.

Shapes of Molecules.   When two atoms are joined in a molecule, they are always arranged in a straight line. There is no other shape for them to take. An example is H_2, the molecule made of two hydrogen atoms joined together in a covalent bond.

The case of molecules made of three or more atoms is more complex. Think about a molecule having one atom in the center, or a central atom, with other atoms arranged around it. The shape of this molecule will be caused by the force pushing electron pairs away from each other. Since all electrons have like negative charges, all electron pairs repel, or push away from, each other. In the methane molecule (CH_4), four identical electron pairs are shared between the C atom and the four H atoms. Each shared electron pair, or covalent bond, attaches an H atom to the C atom. At the same time, each electron pair repels the other three electron pairs. This gives the molecule a tetrahedral shape, a solid with four faces. The C atom is the central atom, and the H atoms are at the four corners. See Fig. 21-1. Each covalent bond is the same and is very stable. This makes methane a nonpolar molecule. It also explains why methane does not react easily with other substances.

Scientists can report the shape of molecules by giving in degrees the **bond angle**, or the angle between two covalent bonds. The H-C-H in methane forms a bond angle of almost 109.5°.

Fig. 21-1

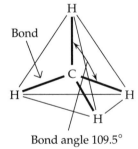

Bond angle 109.5°

 1.   How many covalent bonds make up a bond angle? _____

Like the carbon in methane, the nitrogen atom (N) in ammonia (NH_3) is also the central atom. The N atom has four electron pairs, too. But only three of its electron pairs are shared. These make up three covalent bonds between the N atom and each H atom. The other electron pair is *unshared* (not shared). See Fig. 21-2.

Fig. 21-2   Fig. 21-3

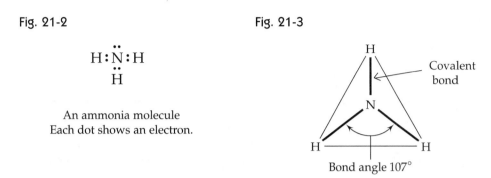

An ammonia molecule
Each dot shows an electron.

Bond angle 107°

Scientists know that unshared electron pairs strongly repel shared electron pairs. This makes the H-N-H bond angle a little smaller than the H-C-H bond angle in methane. Ammonia's bond angle is 107°. See Fig. 21-3. This molecule has a pyramid shape.

Three-Atom Molecules. Sometimes three-atom molecules can be arranged in a straight line. Carbon dioxide (CO_2) is such a molecule. The C atom is the central atom, with an O atom on each side. The C atom shares two electron pairs with the O atoms. Each O atom pulls the electrons more strongly than the C atom, forming polar covalent bonds. But because of their straight line arrangement, the polar bonds cancel each other out (like a tug-of-war between two equally strong people). So carbon dioxide is a nonpolar molecule.

The same is not true for another three-atom molecule: water (H_2O). The H_2O molecule does not look like CO_2. In H_2O, the O atom is the central atom and has four electron pairs. See Fig. 21-4.

Fig. 21-4   Fig. 21-5

A water molecule   Bond angle 104.5°

The O atom shares two of the pairs with H atoms. Two of the pairs are not shared. These unshared pairs repel each other, and they repel the shared electron pairs very strongly. This causes the H-O-H bond angle to be 104.5°. See Fig. 21-5. Scientists say the water molecule is bent, or has an angular shape. This bent shape is important in making water a polar molecule. The O atom in the bent molecule has a negative charge, and the H atoms have positive charges. This, recall from Lesson 20, is why many substances dissolve easily in water.

Molecules have *many* different shapes. Yet all these differences result from unequal repulsion between shared and unshared electron pairs.

Fig. 21-6 presents shapes of various molecules.

Fig. 21-6

Some Molecular Shapes

Molecule	Unshared electron pairs	Bond angle (°)	Shape
CH_4	none	109.5	
NH_3	1	107	
H_2O	2	105	
CO_2	none	180	

Check Your Understanding

Complete the sentences by circling the correct underlined term.

2. A bond angle is formed between the covalent bonds linking <u>two / three</u> atoms.

3. Electron pairs in a molecule are as <u>close to / far apart from</u> each other as possible.

4. The presence of an unshared electron pair in a molecule causes the bond angles to be slightly <u>larger / smaller</u> than they otherwise would be.

5. A molecule containing identical covalent bonds between its atoms, with no unshared electron pairs, is <u>polar / nonpolar</u>.

6. A three-atom molecule with a bent shape is <u>CO_2 / H_2O</u>.

Fig. 21-7 shows the shape of four molecules. Write the letter of the molecule with the correct shape next to its formula.

7. H_2O _____   9. NH_3 _____

8. CH_4 _____   10. CO_2 _____

Fig. 21-7

a   b   c   d

11. What is the bond angle between the atoms in a molecule of oxygen gas (O_2)? Explain your answer. _____

_____

12. What is the shape of a molecule made of two atoms? _____

_____

13. Explain why the shape of a molecule depends on the electron pairs around the central atom. _____

_____

14. Why is the bond angle in ammonia (NH_3) smaller than the bond angle in methane (CH_4)? _____

_____

_____

_____

15. Why is the bond angle in water (H_2O) smaller than the bond angle in ammonia (NH_3)? _____

_____

16. The bent shape of the water molecule (H_2O) is a reason many substances dissolve well in it. Explain why. _____

_____

22

Formulas of Ionic Compounds

Key Words

formula of an ionic compound: chemical formula that tells the kinds and relative numbers of atoms of each kind in an ionic compound

radical: group of atoms of different elements that act together to form an ion

KEY IDEAS

The chemical formula of an ionic compound shows the kinds of atoms and the relative numbers of atoms of each kind. To write a formula of an ionic compound, you need to know the charge on each ion in the compound.

Dental hygienists help people keep their teeth healthy. Hygienists often tell their patients to use fluoride toothpaste to prevent cavities. Fluoride is the anion of the element fluorine. The fluoride ions actually become part of your teeth and make them stronger.

Understanding Chemical Formulas. An easy way to show a chemical compound is to write a chemical formula. The **formula of an ionic compound** uses symbols of elements to show the kinds of atoms in an ionic compound. NaCl is one example. The formula also shows the relative numbers of each kind of atom in a sample of the compound. The formula NaCl shows that there is one Na atom for every Cl atom in a sample of this compound.

The formula $Ca_{10}(PO_4)_6(OH)_2$ describes the crystal substance that makes up your teeth. You can use the periodic table in Lesson 18 to figure out what this formula means. A subscript, or number at the lower right of a symbol, tells how many atoms of that element there are compared to other atoms in the compound. The Ca_{10} shows there are 10 atoms of the element calcium present. The (PO_4) stands for 1 atom of phosphorus bonded to 4 atoms of oxygen. PO_4 is a **radical** (RAD-ih-kuhl), or a group of atoms that bond together. A radical has electric charge and acts like an ion. There are 6 (PO_4) radicals in this formula. The (OH), or hydroxide, is another radical. This formula has 2 hydroxide radicals.

1. Explain what the symbols and numbers mean in the chemical formula of an ionic compound. _____

_____

_____

Writing Formulas. Many formulas are much simpler than the one just discussed. To figure out the formula for an ionic compound, you need to know the charge on each ion. That is, you need to know how many electrons each atom loses or gains when it fills its outer energy level. This is the electric charge—either positive or negative—of the ion. Look at the charges on several ions shown in Fig. 22-1. The positive and negative charges in a compound must add up to zero.

Here's an example: Write the formula for the ionic compound formed by the reaction of sodium and fluorine.

First, write the symbols for the elements in the compound. The symbol for the cation (positively charged ion) goes first. Then write the symbol for the anion (negatively charged ion).

sodium   fluorine

Na   F

Second, in Fig. 22-1, find the electric charge for each ion. Write the charges as superscripts at the upper right of each symbol. The Na atom forms a cation with charge of 1+. The F atom forms an anion with charge of 1–.

Na^{1+} F^{1-}

Third, crisscross the numbers from top to bottom. But do not write the + or – charges. See Fig. 22-2.

Fig. 22-2

$Na_{1}F_{1}$

When the number at the lower right, which is a subscript, is 1, don't use it. Also, when all the subscripts are the same in an ionic compound, don't use them. This process gives the correct formula: NaF. Notice that one positive ion and one negative ion make an electrically neutral compound.

To name a compound like this, use the name of the first element (such as, sodium). Then use the name of the second element but end it with -ide (fluorine becomes fluoride). The fluoride ions in dental rinses that help keep your teeth healthy come from NaF. In fact, two fluoride ions replace the two OH radicals in the tooth crystal, forming $Ca_{10}(PO_4)_6F_2$. This can take place because both the fluoride ion and the hydroxyl radical have an electric charge of 1–.

Fig. 22-1

Charges of Ions

Positive Ions

Na^{1+}	Ba^{2+}
K^{1+}	Ca^{2+}
Ag^{1+}	Mg^{2+}
	Al^{3+}

Negative Ions

Br^{1-}	O^{2-}
Cl^{1-}	S^{2-}
F^{1-}	
OH^{1-}	
I^{1-}	

2. Using Fig. 22-1, write the formula for the compound formed by hydrogen and chlorine. _____

Name the compound. _____

You know that ionic compounds don't form molecules. So the formula NaF simply shows that any sample of sodium fluoride will contain one atom of sodium for every one atom of fluorine.

Fig. 22-3 describes the different types of bonds atoms form.

Fig. 22-3   **Steps to Write Formulas for Ionic Compound**

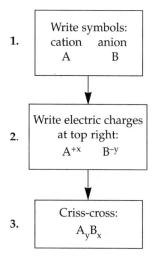

Check Your Understanding

Fill in the information needed to show the chemical formula for a compound made of magnesium and fluorine.

3. Write the symbols for the elements: first cation, then anion.

   Mg _____

4. Look at Fig. 22-1 and then write in the electrical charges for these ions.

   Mg _____   F _____

5. Crisscross the numbers to below the symbols, without using 1, +, or –.

   Mg _____   F _____

6. This ionic compound is called magnesium _____.

7. Any sample of this compound will contain _____ atom(s) of magnesium for every _____ atom(s) of fluorine.

8. What is the chemical formula for an ionic compound? _____

   _____

   _____

9. What do the subscripts, or numbers below the symbols, mean in the chemical formula for an ionic compound? _____

   _____

10. Define a radical and give an example of one. _____

    _____

    _____

11. Using Fig. 22-1, write the formula for the compound formed by calcium and chlorine. _____

    Name the compound. _____

12. Explain why the formula KBr does not tell the kinds and numbers of atoms in a molecule of potassium bromide. What does it tell?

    _____

    _____

    _____

13. Using Fig. 22-1, decide if Cl_2Ba is a correct formula. Explain your answer. If it is not, what is the correct formula? Name the compound.

    _____

    _____

    _____

14. Why do chemists use chemical formulas? _____

    _____

23 Formulas of Covalent Compounds

KEY IDEAS

Molecular formulas are chemical formulas used to show the kinds and numbers of atoms in a molecule. Since the same two elements may combine to form two or more compounds, their molecular formulas must be different.

John Dalton, who developed the atomic theory in the early 1800s, thought a water molecule contained an equal number of hydrogen and oxygen atoms (HO). Today scientists use the molecular formula H_2O to show there are actually two atoms of hydrogen and one atom of oxygen in each water molecule.

Understanding Molecular Formulas. A formula show the kinds of atoms in a sample of a compound. A formula of an ionic compound shows the relative numbers of ions in a sample of an ionic compound, but not their actual numbers. However, because covalent compounds form molecules, chemical formulas give different information about them. A **molecular formula** (moh-LEHK-yoo-luhr) shows the kinds and actual numbers of atoms in a molecule of a covalent compound.

 1. What information does the molecular formula PCl_5 give you?

_____

_____

Some molecules are formed by two atoms of the same element. For instance, hydrogen gas is made of molecules that contain two H atoms. Its molecular formula is simple: H_2.

Writing Formulas. For the formulas of some covalent compounds, you can use almost the same rules as for ionic formulas. Take the molecule formed by hydrogen (H) and fluorine (F) as an example.

Write down the symbols. Put first the element that holds the electrons in the covalent bond less tightly. This one is farther left in the periodic table. Then write the symbol for the element that holds the electron pair more tightly (the one farther right in the table).

$$\text{hydrogen} \qquad \text{fluorine}$$
$$\text{H} \qquad\qquad \text{F}$$

Place at the upper right the charges that each atom would have if it did form an ion (1+ for H and 1– for F). These are not actual electric charges because these atoms don't form ions. Instead, they are called apparent charges.

$$H^{1+} \qquad\qquad F^{1-}$$

Now crisscross the apparent charges. See Fig. 23-1.

Fig. 23-1

$$H_1 \, F_1$$

This gives the molecular formula of HF. Remember, don't use the number one in a subscript. This molecule is called hydrogen fluoride.

The apparent charges in a molecule must add up to zero. This is easy to see for HF (1+ + 1– = 0). Now consider water (H_2O), in which the O atom has an apparent charge of 2–. Each H atom has an apparent charge of 1+. Here the two charges ($2 \times 1+ = 2+$) for H plus the 2– charge for O add up to 0. This forms a neutral H_2O molecule.

Covalent Compounds. Most ionic compounds are formed by a cation from the left side of the periodic table and an anion from the right side. But some covalent compounds contain two elements from the right side, such as sulfur dioxide (SO_2). Both S and O tend to hold their outer energy level electrons tightly. So first write down the element that holds the electron pair less tightly (this one is farther left in the periodic table). Give it a positive apparent charge. The second element (to the right in the table) keeps its apparent negative charge. Follow the crisscross method from here.

Another tricky part of formulas for molecular compounds is that sometimes atoms of an element can have more than one apparent charge. For example, nitrogen (N) and oxygen (O) can form both N_2O and NO. In N_2O, the N atom has an apparent charge of 1+. Because there are two N atoms, this gives a total positive charge of 2 ($2 \times 1+ = 2+$). This balances out the apparent negative charge of the O, which is 2–. But in NO, the N atom's apparent charge is 2+. This 2+ added to the 2– for O equals 0, making the molecule electrically neutral.

Of course, because N_2O and NO are different compounds, they also have different names. In cases like this, where both elements are from the right side of the periodic table, chemists use the prefixes shown in Fig. 23-2. So N_2O is known as dinitrogen oxide, and NO is called nitrogen monoxide. (Just remember not to use *mono-* as part of the first word.)

Fig. 23-2

Prefix	Meaning
mono–	1
di–	2
tri–	3
tetra–	4
penta–	5

2. Using Fig. 23-2, name the molecule with the formula N_2O_3.

Fig. 23-3 summarizes the steps used to write formulas for covalent compounds.

Fig. 23-3

Steps to Write Formulas for Covalent Compounds

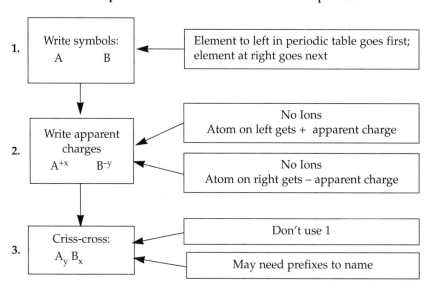

Check Your Understanding

Answer the following questions about one of the covalent compounds formed by carbon and oxygen: CO.

3. How many atoms of carbon are in a molecule of CO? Of oxygen?

    _____

4. Which element holds electrons more tightly in CO? Explain.

    _____

    _____

5. What is the apparent charge of the carbon atom in CO? _____

6. What is the apparent charge of the oxygen atom in CO? _____

7. What is the name of CO? _____

8. Explain the difference between a molecular formula and the chemical formula of an ionic compound. _____

_____

_____

_____

9. What information does the molecular formula HNO_2 give you?

_____

_____

10. Explain how one element can combine with different numbers of atoms of another element to form different molecular compounds.

_____

_____

_____

11. Name the following covalent compounds.

    (a) CO_2 _____

    (b) SO_3 _____

    (c) CCl_4 _____

    (d) N_2O_5 _____

    (e) H_2S _____

12. Here are a few apparent charges for atoms in various covalent compounds: C — 4+; P — 3+; S — 2-, 4+. Give the apparent charge on each of the underlined atoms.

    (a) $\underline{C}O_2$ _____   **(c)** $\underline{H}_2S$ _____

    (b) $P\underline{Cl}_3$ _____   **(d)** $\underline{H}_2O$ _____

13. Why do the atoms in molecules have apparent electric charges and not actual electric charges? _____

_____

_____

Unit 4 Review

Summary

- Chemical bonds are forces that hold atoms together in a compound.

- In ionic bonds, electrons are transferred from the outer energy level of one atom to the outer energy level of another atom.

- Atoms that lose electrons in an ionic bond form positively charged ions called cations.

- Atoms that gain electrons in an ionic bond form negatively charged ions called anions.

- Some ionic compounds form crystals that can conduct an electric current when dissolved in water.

- In covalent bonds, an electron pair is shared between atoms.

- Atoms held together by covalent bonds form a molecule.

- Many covalent compounds are poor conductors of electricity, although polar covalent bonds give the atoms within a molecule some electric charge.

- The shapes of molecules depend on electron pairs repelling, or pushing away from, each other.

- Molecular shape can be measured by the bond angle, or the angle formed by two covalent bonds.

- Chemical formulas for ionic compounds show the kinds of atoms present, but not their actual number.

- Radicals are groups of atoms joined by covalent bonds that together act like ions.

- Molecular formulas show the kinds and actual numbers of atoms in molecules of covalent compounds.

For Your Portfolio

1. Use toothpicks and different-sized styrofoam balls to make some molecular models. First try a molecule of methane. Remember to make the bond angles equal. Then make models of the other molecular shapes you learned in Lesson 21.

2. Use toothpicks and gumdrops of two colors to make a model of a sodium chloride (NaCl) crystal. Remember there is one Na ion for each Cl ion.

3. The Stock system is a way to name compounds in which the atoms can form more than one kind of ion. Learn about the Stock system by doing some research in the library. Then name the following compounds: $FeCl_2$ and $FeCl_3$.

Unit 4 PRACTICE TEST

Column I contains definitions of some terms. Column II contains the terms being defined. On each line, write the letter of the term from Column II that goes with each definition.

	Column I	Column II
_____	1. group of atoms that together act like a single anion	**a.** ion
_____	2. force that holds atoms together in a compound	**b.** radical
_____	3. group of atoms held together by covalent bonds	**c.** molecule
_____	4. solid in which particles are in a regular, repeating pattern	**d.** crystal
_____	5. atom that develops an electric charge when it loses or gains electrons	**e.** chemical bond

Write the formula for the compound formed by each of the following pairs of atoms.

6. sodium and oxygen _____

7. fluorine and fluorine _____

8. barium and chlorine _____

9. hydrogen and bromine _____

10. calcium and iodine _____

Choose the term that makes each sentence correct. Then write the letter of the term in the space at the left.

_____ 11. A molecule of CH_4 has

    a. one chemical bond.   **c.** a bond angle of $0°$.
    b. four chemical bonds.   **d.** a bond angle of $180°$.

_____ 12. The transfer of electrons from one atom to another forms a(n)

    a. polar covalent compound.   **c.** ionic compound.
    b. non-polar covalent compound.   **d.** non-ionic compound.

_____ 13. When salt dissolves in water, which of the following happens?

    a. Na^+ ions are pulled toward the oxygen atom of the water molecules
    b. Na^+ ions replace the hydrogen atoms in the water molecules
    c. the Cl^- ions become neutral
    d. the Cl^- ions combine with the hydrogen atoms

_____ 14. In a molecule of H_3PO_4, there are

    a. 3 H atoms.   **c.** 4 O atoms.
    b. 3 P atoms.   **d.** 4 PO radicals.

_____ 15. A sulfide ion can be either S^{2-} or S^{4+}. A sodium ion is always Na^{1+}. The formula of sodium sulfide is

    a. SNa_4.   **b.** S_4Na.   **c.** NaS_2.   **d.** Na_2S.

Give a brief answer for each of the following.

16. What is the difference between an ionic bond and a covalent bond? _____

   _____

   _____

   _____

17. What do the symbols and numbers in a chemical formula stand for? _____

   _____

   _____

   _____

Answer one of the two questions.

18 a. Explain what is meant by a polar covalent bond. How is it different from a nonpolar covalent bond? From an ionic bond? What important property does having polar bonds give to a compound? Name two molecules with this type of bond.

   b. Explain why atoms that give up electrons form positively charged ions. Explain why atoms that gain electrons form negatively charged ions. Give some examples of each type of ion.

Quantitative Chemistry

Did you ever bake bread "from scratch?" If so, you probably used a recipe. Could you have made the bread just by knowing that it contains flour, shortening, salt, sugar, milk, water, and yeast? Probably not. You would need a recipe to tell you how much of each ingredient, how to combine them, how long to let the dough rise, what kind of pan to put the dough in, and what temperature to use to bake the bread.

You know that chemical changes are involved in baking bread. The properties of the raw materials you put into the dough are different from those of the finished product. You can compare a recipe to a balanced chemical equation. This recipe tells you the correct proportions of the ingredients to make two loaves of bread in two loaf pans.

The recipe for bread that tells how much of each ingredient you need could be written like a chemical equation: 6 cups white flour + 2 tablespoons shortening + 2 $\frac{1}{2}$ teaspoons salt + 2 tablespoons sugar + 1 cup milk + 1 $\frac{1}{4}$ cups water + 1 package yeast. If you leave out one of the ingredients, say the yeast, the bread will not rise. You can't eat the bread. It is the same in a chemical reaction. If you leave out a reactant or use the wrong amount, the product you expect will probably not form.

6.02×10^{23}

6.02×10^{23}

6.02×10^{23}
6.02×10^{23}
6.02×10^{23}
6.02×10^{23}
6.02×10^{23}
6.02×10^{23}

6.02×10^{23}

6.02×10^{23}

6.02×10^{23}

6.02×10^{23}

6.02×10^{23}
6.02×10^{23}
6.02×10^{23}
6.02×10^{23}
6.02×10^{23}
6.02×10^{23}
6.02×10^{23}
6.02×10^{23}

Lesson 24 The Mole

Key Words

coefficient: number in front of a formula

formula mass: sum of the atomic masses of all atoms indicated in the formula of a substance

gram formula mass: formula mass expressed in grams of a substance

gram atomic mass: atomic mass of an element expressed in grams

mole: the number 6.02×10^{23}

molar mass: mass of one mole of an element or compound

KEY IDEAS

Mole is a word for a specific number. Compare the word mole to the word dozen. One dozen means 12. One mole means 6.02×10^{23}. One mole of a substance has a mass equal to the mass of 6.02×10^{23} of its molecules or atoms. There are 6.02×10^{23} molecules or atoms in one mole of a substance.

Fig. 24-1

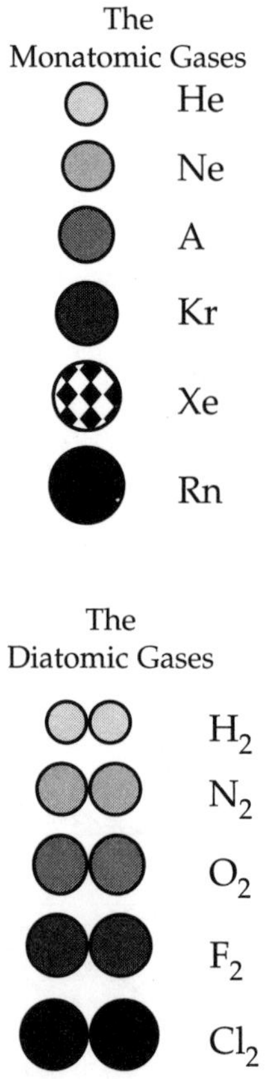
The Monatomic Gases
He
Ne
A
Kr
Xe
Rn
The Diatomic Gases
H_2
N_2
O_2
F_2
Cl_2

Formulas for Gaseous Elements. Eleven of the elements exist as molecules of gases. As shown in Fig. 24-1, six of these elements have monatomic, or one-atom, molecules. The other five have diatomic, or two-atom, molecules. The subscript 2 shows that there are two atoms in each diatomic molecule.

Formula Units. The formula of a compound is called a formula unit. Look at examples of the number of atoms in formula units.

Example 1: Hydrogen—H_2. In one formula unit of H_2, there are two atoms of hydrogen. Two or more formula units are written using a **coefficient** (koh-uh-FIHSH-uhnt), which is a number in front of the formula. For example, in $2H_2$, there are two formula units, that have twice as many atoms as there are in one formula unit.

Example 2: Magnesium hydroxide—$Mg(OH)_2$. The subscript 2 refers to both atoms enclosed by parentheses. Thus, in one formula unit, there are one atom of magnesium, two atoms of oxygen, and two atoms of hydrogen.

✓ **1. How many atoms of each kind are there in each of the following?**
(a) O_2 (b) $3O_2$ (c) $4CO_2$ (d) one formula unit of magnesium phosphate—$Mg_3(PO_4)_2$

(a) ______________________ (b) ______________________

(c) ______________________ (d) ______________________

Formula Masses. The **formula mass** (FAWR-myoo-luh) of a substance is the sum of the atomic masses of all atoms in one formula unit of the substance. Remember that atomic masses appear in the periodic table. Look at the examples.

Example 1: Water—H_2O. The atomic masses of the 2 hydrogen atoms are $2 \times 1.01 = 2.02$. The atomic mass of the 1 oxygen atom is 16.00. So the formula mass of water is $2.02 + 16.00 = 18.02$.

Example 2: Magnesium hydroxide—$Mg(OH)_2$. The table shows how to calculate the formula mass, which is 58.33.

Element	Atomic mass (from the periodic table)	Number of atoms shown by the formula	Mass of all atoms shown by the formula
Magnesium	24.31	1	24.31
Oxygen	16.00	2	32.00
Hydrogen	1.01	2	2.02
TOTAL			58.33

  2. **Find the formula masses of the following substances:**
 (a) **nitrogen gas—N_2** (b) **CO_2**
 (c) **magnesium phosphate—$Mg_3(PO_4)_2$**
 (a) _____ (b) _____ (c) _____

Gram Formula Mass. The **gram formula mass** is the formula mass of a substance expressed in grams. Gram formula mass has a number and a unit. The number is the formula mass of the substance and the unit is grams. The gram formula mass of water is 18.02 grams.

Gram Atomic Mass. The **gram atomic mass** (uh-TAHM-ihk) is the atomic mass of an element with the unit *grams* following it. For example, the atomic mass of sodium is 23.0, or 23.0 atomic mass units. The gram atomic mass of sodium is 23.0 grams.

The Mole. Mole (abbreviated *mol*) is a word for a specific number. One **mole** (mohl) is 6.02×10^{23} atoms or molecules of a substance.

Example 1: A mole of oxygen atoms is 6.02×10^{23} oxygen atoms. The number 6.02×10^{23} is called the Avogadro number.

Example 2: A mole of oxygen molecules is 6.02×10^{23} oxygen molecules. There are two moles of oxygen atoms in a mole of oxygen gas.

Two moles is $2 \times (6.02 \times 10^{23}) = 12.04 \times 10^{23}$. Notice that when you multiply by 2, the 2 multiplies only the 6.02.

3. **Suppose you have 3 moles of carbon atoms. How many actual atoms do you have?** _____

A mole of an element is 6.02×10^{23} atoms of that element. A mole of a compound is 6.02×10^{23} formula units of that compound. The following table shows this concept.

Compound	Number of atoms in one formula unit	Number of atoms in one mole of compound
H_2O	2 H atoms	2 moles of H atoms
	1 O atom	1 mole of O atoms
$Mg(OH)_2$	1 Mg atom	1 mole of Mg atoms
	2 O atoms	2 moles of O atoms
	2 H atoms	2 moles of H atoms

 4. **In terms of moles, how many atoms of each element are present in 1 mole of $Mg_3(PO_4)_2$?** _____

_____

Relation Between Mole and Mass. One mole of a substance has a mass equal to the mass of 6.02×10^{23} of its molecules or atoms. The gram formula mass of a substance and a mole of the substance are related. Suppose that you have a sample of 6.02×10^{23} formula units of a compound, or a 1-mole sample. The mass of the sample is equal to the gram formula mass. The gram formula mass is also called the **molar mass** (MOH-luhr). When you say "one mole," you are talking about 6.02×10^{23} particles. One mole of water is 6.02×10^{23} molecules of water. When you say "molar mass," you are talking about the mass of a number of particles equal to one mole.

Example: The gram formula mass of water is 18.02 grams. Therefore, the molar mass of water is 18.02 grams. One mole of water has a mass of 18.02 grams.

Fig. 24-2 shows how to find the formula unit, number of atoms of each element in a mole, formula mass, gram formula mass, and molar mass for water—H_2O. Note that gram formula mass equals molar mass.

Fig. 24-2

Formula	H_2O	
One formula unit	2 hydrogen atoms	1 oxygen atom
One mole	2 moles of hydrogen atoms	1 mole of oxygen atoms
Formula mass	18.02	$1.01 \times 2 = 2.02$
		$= 16.00$
		18.02
Gram formula mass	18.02 grams	
Molar mass	18.02 grams	

Write the missing term in each blank.

5. There are two atoms in a _____ molecule.

6. The 3 in front of 3NaCl is called a _____.

7. In $(NH_4)_2SO_4$, the _____ 2 applies to both the N and H atoms.

8. A substance's formula mass is the sum of all the _____ masses of the atoms indicated by its formula.

9. If the formula mass of a substance is 10.00, then its _____ mass is 10.0 grams.

10. The molar mass of a substance is the same as its _____ mass.

---

11. Draw a picture showing the difference between a diatomic molecule and a monatomic molecule.

What Do You Know?

12. How is the formula for hydrogen gas different from the formula for helium gas? _____

_____

13. For potassium chlorate—$KClO_3$, **(a)** how many atoms of each element are represented by one formula unit? **(b)** By three formula units?

    (a) _____

    (b) _____

You may use the periodic table to answer the following questions.

14. What is the formula mass of chloroform—$CHCl_3$? _____

_____

15. **(a)** What is the formula mass of potassium chlorate—$KClO_3$? **(b)** What is its molar mass?

    (a) _____ **(b)** _____

Percentage Composition and Empirical Formulas

percentage composition:	percent of each element in a compound
empirical formula:	formula that shows the simplest possible whole-number subscripts in the formula of a compound
molecular formula:	formula that shows the number of atoms of each element in one molecule

KEY IDEAS

From the known masses of each element in a sample of a compound, you can calculate the percentage composition and the empirical formula of the compound. If you know the formula mass, you can also calculate the molecular formula.

Most of the metals you use every day come from minerals found in the earth. Chemists need to know how much of each element there is in a particular mineral. The percentage composition of a mineral gives chemists this information.

Percentage Composition. If you know the mass of each element in a sample of a compound, you can calculate its percentage composition. The **percentage composition** (per-SEHNT-uhj kahm-puh-ZIH-shuhn) is the percentage of each element in a compound. Suppose a sample of a compound contains 7.85 g nitrogen and 17.92 g oxygen. To find its percentage composition, first divide the mass of each element in the sample by the mass of the sample. In this sample, the mass is 7.85 g + 17.92 g or 25.77 g. Then multiply each answer by 100 to convert the decimal to a percentage.

$$\frac{7.85 \text{ g nitrogen}}{7.85 \text{ g} + 17.92 \text{ g}} \times 100 = 30.46\% \text{ N}$$

$$\frac{17.92 \text{ g oxygen}}{7.85 \text{ g} + 17.92 \text{ g}} \times 100 = 69.54\% \text{ O}$$

1. A compound contains two elements, sulfur and oxygen. In a sample of the compound, the mass of sulfur is 6.41 grams. The mass of the oxygen is 9.60 grams. What is the percentage composition of the compound? Show your work. _____

Empirical Formulas. An **empirical formula** (ehm-PIHR-ih-kuhl FAWR-myoo-luh) has subscripts that cannot be divided evenly by a whole number other than one. (A subscript is the number at the lower right of a symbol.) In a compound with two or more elements, the empirical formula shows the relative number of atoms of each element. With that in mind, is H_2O_2 an empirical formula? Let's test the formula to see.

Element	Subscript	
hydrogen	2	$\dfrac{2}{2} = \dfrac{1}{1}$
oxygen	2	

The formula H_2O_2 shows that there are two hydrogen atoms for every two oxygen atoms. The subscript ratio in this formula is 2 to 2, which is the same as the ratio of 1 to 1. This means that there is 1 hydrogen atom for each oxygen atom in all samples of hydrogen peroxide. Therefore, the empirical formula of hydrogen peroxide is H_1O_1, or HO. The empirical formula shows nothing about the total number of atoms in a molecule.

 2. What is the empirical formula of each of the following?
   (a) C_2H_6 _____   (b) C_2H_4 _____
   (c) C_2H_2 _____   (d) NO_2 _____

Empirical Formulas from Percentage Composition. If you know the percentage composition of a compound, you can find its empirical formula. The percentages of the nitrogen-oxygen compound calculated on page 120 are 30.46% nitrogen and 69.54% oxygen. To find the empirical formula of the compound, follow these steps:

Step 1: *Assume that you have a 100-gram sample.*
That would mean 30.46 grams N and 69.54 grams O.

Step 2: *Find the number of moles of each element in the sample.* To do this, divide each mass in grams by its molar mass. Use the periodic table to find the molar mass and any other information you may need.

$$\frac{30.46 \text{ g nitrogen}}{14.01 \text{ g/mol}} = 2.17 \text{ moles nitrogen}$$

$$\frac{69.54 \text{ g oxygen}}{16.00 \text{ g/mol}} = 4.34 \text{ moles oxygen}$$

Step 3: *Convert the number of moles from Step 2 into the simplest whole-number ratio.* To do this, divide the smaller number of moles found in Step 2 into each number. The number of moles of nitrogen—2.17—is the smaller number. Therefore, divide each number of moles by 2.17.

$$\frac{2.17 \text{ moles nitrogen}}{2.17} = 1 \text{ mole nitrogen}$$

$$\frac{4.34 \text{ moles oxygen}}{2.17} = 2 \text{ moles oxygen}$$

These numbers, 1 and 2, are the subscripts for the empirical formula. Because 2 moles is twice as many atoms as 1 mole, the empirical formula is N_1O_2, or NO_2.

3. A compound is 46.68% nitrogen and 53.32% oxygen. Find its empirical formula. Show all work. _____

_____

Molecular Formulas. The **molecular formula** (moh-LEHK-yoo-luhr) of a compound shows the number of atoms of each element in one molecule of the compound. The molecular formula is sometimes the same as the empirical formula. This is true of water, whose molecular formula is H_2O. The subscripts 2 and 1 cannot be divided by a whole number (other than 1) to produce a ratio simpler than a 2 to 1 ratio. Therefore, H_2O is both the molecular formula and the empirical formula of water.

4. Is C_6H_6 a molecular formula or an empirical formula? Explain.

_____

_____

Determining the Molecular Formula. Suppose you know both the empirical formula of a compound and the formula mass of the compound. From this information, you can find the molecular formula of the compound.

Example: The empirical formula of a compound is NO_2. Its formula mass is 92.0. Use these steps to find its molecular formula.

Step 1: *Determine the formula mass for the empirical formula.* The formula mass of NO_2 is $14.0 + 32.0 = 46.0$.

Step 2: *Make a ratio of the given formula mass to the empirical formula mass. Use this ratio to adjust the subscripts.* The ratio of the given formula mass to the empirical formula mass is 92 to 46, or 2 to 1. Therefore, the subscripts in the empirical formula NO_2 must be doubled. Doubling them gives N_2O_4. The molecular formula of the compound is therefore N_2O_4.

Fig. 25-1 shows how to find the empirical formula of a compound.

Fig. 25-1

Determining the Empirical Formula of a Compound

(GIVEN: You know the masses of element R and element S in a compound of the two elements.)

Mass of element R in compound	Mass of element S in compound

Dividing mass of R by mass of compound gives

Dividing mass of S by mass of compound gives

Percent of element R in compound	Percent of element S in compound

Dividing percent R by its molar mass gives

Dividing percent S by its molar mass gives

Number of moles of atoms of R	Number of moles of atoms of S

Dividing each number of moles of atoms by the smaller number of moles gives the

Smallest whole numbers, which are the subscripts in the empirical formula

Check Your Understanding

On each blank, write the missing term.

5. A(n) _____ formula tells the relative number of atoms of each element in a compound but not the number of atoms in a molecule.

6. The _____ formula of a compound can be found if you know the masses of each element in a sample.

7. A formula is a(n) _____ formula if its subscripts cannot be divided evenly by a whole number other than one.

8. Which of the following formulas is an empirical formula: N_2O_4, Fe_2O_3, H_2O_2, or Mg_2Cl_4? Explain. _____

_____

9. In a compound of sulfur and oxygen, the mass of sulfur is 3.2 g. The mass of oxygen is 4.8 g. What is the percentage composition of the compound? Show your work.

_____

10. The empirical formula of a compound is CH_2. Its formula mass is 42.09. What is the compound's molecular formula?

_____

Chemical Equations

KEY IDEAS

Chemical equations are used to describe what happens during a chemical reaction. Equations must contain the correct symbols and formulas for the reactants and the products. Coefficients must be added to show that atoms are conserved.

Fig. 26-1

Before reaction

H₂ molecule   O₂ molecule

After reaction

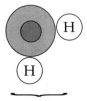

H₂O molecule

Chemical Equations. Chemical equations describe chemical changes, or reactions. For example, hydrogen and oxygen react to form water. Fig. 26-1 shows hydrogen (H_2) and oxygen (O_2) molecules before they react. It also shows a water (H_2O) molecule produced by the reaction.

Fig. 26-1 shows the number of atoms in hydrogen and oxygen molecules. There are two hydrogen atoms in an H_2 molecule and two oxygen atoms in an O_2 molecule. After reaction, there are two hydrogen atoms but only one oxygen atom in one molecule of H_2O. The second oxygen atom that was present before reaction is missing. So the picture equation needs to be balanced.

A Balanced Equation. To balance an equation means to make the number of atoms of each element before and after the reaction equal. You can balance the picture equation by adding another H_2 molecule before the reaction and another H_2O molecule after the reaction. Fig. 26-2 shows the result of adding these molecules. The picture equation now balances.

The formula equation shown in Fig. 26-2 is as follows:

$$2H_2 + O_2 \longrightarrow 2H_2O$$

In words, this equation says, "Two hydrogen molecules react with one oxygen molecule to produce two water molecules." Unlike Fig. 26-1, Fig. 26-2 shows that the number of atoms in a reaction is conserved. According to the law of conservation of matter, the total amount of matter before and after a reaction must be the same.

Fig. 26-2

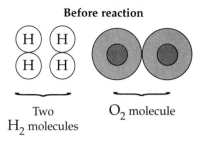

Before reaction

Two H_2 molecules O$_2$ molecule

 1. **Explain why the picture equation balances.**

_____

_____

The plus sign in the equation means "reacts with." The arrow means "to produce." The coefficient 2 before H_2 shows that two molecules of hydrogen react. The coefficient 2 before H_2O shows that two molecules of water are produced.

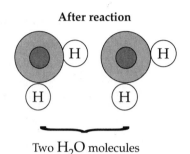

After reaction

 2. **Two formula units of potassium chlorate—$KClO_3$—produce two formula units of potassium chloride—KCl—and three formula units of oxygen. Write a formula equation in place of this word equation.**

_____

Two H_2O molecules

The substances that react with each other are called **reactants** (ree-AK-tuhnts). When H_2 and O_2 react, the hydrogen and oxygen are the reactants. The substance or substances produced during a chemical reaction are called **products** (PRAHD-uhkts). H_2O is the product of the reaction between H_2 and O_2.

Steps in Writing a Balanced Equation

Step 1.   _Write a word equation that states the reactants and the products of the reaction._

Step 2:   _Write the correct symbols and formulas for the reactants and the products. Be sure to use the subscript 2 when writing the formulas of diatomic gases._

Step 3:   _Add coefficients to make the number of atoms of each element equal on both sides of the equation. The coefficients should be the smallest possible whole numbers._

As an example, the steps for the reaction during which hydrogen and oxygen produce water are shown below:

Step 1:   _Write the word equation._ Hydrogen gas and oxygen gas react to produce water.

Step 2:   _Write the correct formulas._ $H_2 + O_2 \longrightarrow H_2O$

Step 3:   _Add the correct coefficients._ $2H_2 + O_2 \longrightarrow 2H_2O$

3. Zinc metal reacts with hydrochloric acid—HCl—to produce zinc chloride—ZnCl$_2$—and hydrogen gas. Write a correctly balanced formula equation for this word equation.

Showing Phase. The substances that take part in chemical reactions may be gases, liquids, or solids. They may also be substances that have been dissolved in water. To show the phases of reactants and products in a chemical equation, you put a letter after the formula. The letter is enclosed in parentheses.

These letters are used to show phase: *g, l, s,* and *aq.* The letter g means the substance is a *gas.* The letters *l* and *s* mean it is a liquid or a solid, respectively. The letters *aq* mean aqueous, or related to water. The word aqueous means that the substance reacts when dissolved in water. These letters have been added to the following equation:

$$2H_2(g) \ + \ O_2(g) \longrightarrow 2H_2O(l)$$

Meanings for Coefficients. So far, the number 2 before H$_2$O has meant 2 molecules. However, coefficients can have other meanings. A coefficient can mean the number of moles of molecules. Thus, the reaction of hydrogen with oxygen can be read, "Two moles of hydrogen react with one mole of oxygen to produce two moles of water."

Fig. 26-3 shows how you can write balanced word and picture equations for the reaction of carbon and oxygen that produces carbon monoxide.

Fig. 26-3

Carbon	reacts with	oxygen	to yield	carbon monoxide	Word equation
●		◯◯		●◯	Picture of reactants and product
C	+	O$_2$	⟶	CO	Formula equation (*un*balanced)
2C	+	O$_2$	⟶	2CO	Formula equation (balanced)
● ●		◯◯ ◯◯	⟶	●◯ ●◯	Balanced picture equation

Write the missing terms in the blanks.

4.   The _____ are the substances that react with each other during a chemical reaction.

5.   A chemical equation is _____ by adding coefficients in front of the symbols or formulas.

6.   A plus sign on the left side of an equation means _____.

7.   The letter g in parentheses following the formula of a substance means the substance is a(n)_____.

8.   The coefficients in an equation can refer to the number of molecules or to the number of _____.

What Do You Know?

9.   When hydrochloric acid is poured over zinc metal, hydrogen gas bubbles from the zinc, and zinc chloride is produced in the acid solution. Write a word equation for this reaction.

   _____

10. Balance the following equations.

   a.   _____ Mg + _____ O$_2$ —> _____ MgO

   b.   _____ Na + _____ H$_2$O —> _____ NaOH + _____ H$_2$

   c.   _____ K + _____ Br$_2$ —> _____ KBr

   d.   _____ Al + _____ Cl$_2$ —> _____ AlCl$_3$

11. Balance the following equations.

   a.   _____ KClO$_3$ —> _____ KCl + _____ O$_2$

   b.   _____ N$_2$ + _____ H$_2$ —> _____ NH$_3$

   c.   _____ Al + _____ NiCl$_2$ —> _____ AlCl$_3$ + _____ Ni

   d.   _____ CH$_4$ + _____ O$_2$ —> _____ CO$_2$ + _____ H$_2$O

12. Write a balanced formula equation for the following reaction. Magnesium metal reacts with silver chloride to produce silver metal and magnesium chloride. _____

Classifying Chemical Reactions

synthesis:	reaction in which two substances combine to form one substance
decomposition:	reaction in which one substance breaks up into two smaller substances
single replacement:	reactions in which one element takes the place of another in a compound
double replacement:	reactions in which elements in two compounds replace each other
precipitate:	solid formed in double replacement reactions when two solutions of ions are mixed

KEY IDEAS

Many common reactions can be classified into four types. These are synthesis, decomposition, single replacement, and double replacement.

Early chemists did not know enough about chemical reactions to be able to classify them. As more chemical reactions were observed, it became possible to classify them according to certain similarities.

Synthesis. In a **synthesis** (SIN-thuh-sihs) reaction, two substances combine to form another substance. The general form of the reaction is $A + B \rightarrow AB$. An example is shown in Equation 1, the reaction between hydrogen and oxygen to produce water.

Equation 1 $\qquad 2H_2 + O_2 \longrightarrow 2H_2O$

Decomposition. A decomposition reaction is the opposite of a synthesis reaction. In a **decomposition** (dee-kahm-puh-ZIHSH-uhn) reaction, one substance breaks apart to form two simpler substances. The general form of the reaction is $AB \rightarrow A + B$. An example is shown in Equation 2, which is the opposite of Equation 1.

Equation 2 $\qquad 2H_2O \longrightarrow 2H_2 + O_2$

FIG. 27-3

1. **Write equations for (a)** the decomposition of carbon monoxide—
CO—to produce carbon—C—and oxygen gas and **(b)** the synthesis
reaction for the formation of carbon monoxide from carbon and
oxygen.

(a) _____

(b) _____

Single Replacement. In a **single replacement** (ree-PLAYS-muhnt) reaction,
one element takes the place of another element in a compound. The general
form of the reaction is A + BC → B + AC. In Equation 3, magnesium replaces
the zinc in zinc sulfate. Fig. 27-1 shows this reaction.

Equation 3 $Mg(s) + ZnSO_4(aq) \longrightarrow Zn(s) + MgSO_4(aq)$

Fig. 27-1

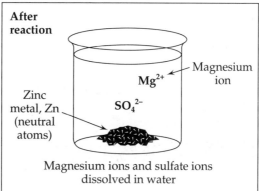

$$Mg + ZnSO_4 \longrightarrow Zn + MgSO_4$$

Fig. 27-2 shows more detail about the reaction in Equation 3. Note that $ZnSO_4$
and $MgSO_4$ are ionic compounds and both are dissolved in water. The (*aq*)
after their formulas in Equation 3 shows that they are both in solution. In
this reaction, neutral magnesium atoms—Mg—become magnesium ions—
Mg^{2+}. For zinc, the reverse happens. Zinc ions—Zn^{2+}—become neutral zinc
atoms—Zn.

Fig. 27-2

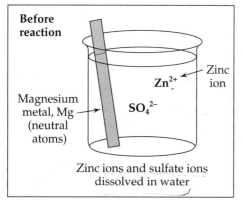

Note that there is no change with the SO_4^{2-} ions.

Equation 4 shows an example of another kind of single replacement reaction.
In this reaction, chlorine (a halogen) replaces bromine (another halogen) in
sodium bromide.

Equation 4 $Cl_2(g) + 2NaBr(aq) \longrightarrow Br_2(l) + 2NaCl(aq)$

In Equations 3 and 4, the element replaced is less active chemically than the element that replaces it. Fig. 27-3 shows the activity series of metals. A metal higher up on the list will replace the ions of a metal lower down on the list. Magnesium is higher than zinc. That's why the reaction shown by Equation 3 takes place. Hydrogen is in the table because the ability of a metal to free hydrogen from an acid is an indication of the metal activity.

The activity of the elements known as halogens is shown in Fig. 27-4. A halogen higher on the list will replace the ions of a halogen lower on the list. Chlorine is higher than bromine. That's why the reaction shown by Equation 4 takes place.

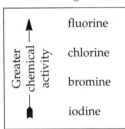

Activity Series

Greater chemical activity

lithium
potassium
barium
calcium
sodium
magnesium
aluminum
zinc
iron
nickel
lead
hydrogen
copper
silver
gold

 2. **Will zinc metal react with silver nitrate—AgNO$_3$—solution in a single replacement reaction? Explain. If it will react, write the equation for the reaction.** _____

_____

Fig. 27-4

Activity Series
of Halogens

Greater chemical activity

fluorine
chlorine
bromine
iodine

Double Replacement. In a **double replacement** reaction, elements in two compounds replace each other. A general form of the reaction is AB + CD → AD + BC. This reaction takes place with the reactants dissolved in water.

There are two common types of double replacement reactions. In one type, cations from one compound combine with anions from another to produce a solid. Equation 5 shows an example.

Equation 5 $\quad$ NaCl(aq) + AgNO$_3$(aq) $\longrightarrow$ NaNO$_3$(aq) + AgCl(s)

The silver and chloride ions form a solid. Equation 5 shows this by the (s) in AgCl(s). Solids formed in double replacement reactions are called **precipitates** (pree-SIHP-uh-tayts). See Fig. 27-5.

Fig. 27-5

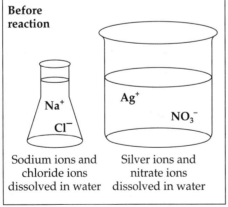

Before reaction

Na$^+$
Cl$^-$

Ag$^+$
NO$_3^-$

Sodium ions and chloride ions dissolved in water

Silver ions and nitrate ions dissolved in water

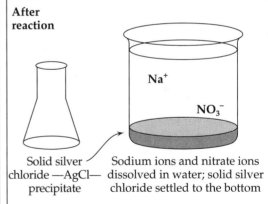

After reaction

Na$^+$
NO$_3^-$

Solid silver chloride —AgCl— precipitate

Sodium ions and nitrate ions dissolved in water; solid silver chloride settled to the bottom

In another type of double replacement reaction, a hydrogen ion from one compound combines with a hydroxide ion from another to form water. Equation 6 shows an example of this kind of reaction.

Equation 6 $\quad$ HCl(aq) + NaOH(aq) $\longrightarrow$ NaCl(aq) + H$_2$O(l)

In the reaction shown in Fig. 27-6, zinc metal—Zn— becomes zinc ions—Zn^{++}— and nickel ions—Ni^{++}—become nickel metal—Ni—. Nothing happens to the sulfate ions—SO_4.

Fig. 27-6

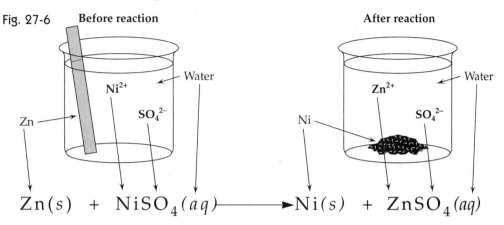

$$Zn(s) \ + \ NiSO_4(aq) \longrightarrow Ni(s) \ + \ ZnSO_4(aq)$$

Check Your Understanding

Write the missing word in each blank.

3. When a substance breaks up into two simpler substances, the reaction is a(n) _____ reaction.

4. A metal that will replace the ions of another metal in water is a more _____ metal than the metal whose ions are replaced.

5. A(n) _____ is the solid sometimes produced when two solutions containing dissolved ions are mixed.

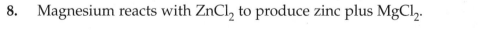

What Do You Know?

Write equations for the reactions in items 6 to 8.

6. Solid $KClO_3$ is heated in a test tube. Crystals of KCl are produced and oxygen gas leaves the test tube. _____

    _____

7. Sodium metal reacts with chlorine gas to produce crystals of sodium chloride. Show phases in your equation.

    _____

8. Magnesium reacts with $ZnCl_2$ to produce zinc plus $MgCl_2$.

    _____

9. Identify the kinds of reactions described in items 6 through 8.

    Item 6 _____   Item 7 _____

    Item 8 _____

Stoichiometry I

Key Words

stoichiometry:	calculation of mass and volume relationships among substances in chemical reactions
mass-mass problem:	a problem in which mass of one substance in a reaction is given and mass of another substance is found

KEY IDEAS

Solving mass-mass problems is a three-step process. You start and end with mass in grams, but in between, convert to and from moles. The key to solving mass-mass problems is the balanced equation.

When foods cook, chemical reactions take place. Recipes tell the cook how much of each ingredient to use in a particular dish. The balanced formula equation tells a chemist how much of each reactant to use to make a product.

Stoichiometry. Stoichiometry (stoi-kee-AHM-uh-tree) deals with mass and volume relationships among substances in chemical reactions. The substance whose mass you are given is called "the given." You use the given mass to find the mass of another substance, called "the unknown." A problem of this kind is called a **mass-mass problem**.

To solve mass-mass problems it is necessary to start with a balanced equation. Look at this general equation:

$$A + B \longrightarrow C + D$$

In this equation, there are four substances—A, B, C, and D. Fig. 28-1 shows some possible combinations of given and unknown substances for the general equation.

Fig. 28-1

Some of the possible combinations of Givens and Unknowns

A	+	B	—>	C	+	D	A	+	B	—>	C	+	D
Given		Unknown									Given		Unknown
A	+	B	—>	C	+	D	A	+	B	—>	C	+	D
Given				Unknown			Unknown						Given

The problem-solving method to be described applies to all equations. It applies to equations with fewer than and more than four substances.

1. What kinds of relationships does stoichiometry deal with?

_____

 2. What do you call the substance whose mass you are to find?

_____

Solving Mass-Mass Problems. Solving mass-mass problems is a three-step process. Each factor shown in Fig. 28-2 is a kind of bridge. Each bridge takes you from the quantity in one box to that in the next. The problem below shows how to use this figure to solve problems.

Fig. 28-2

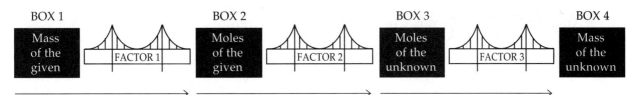

Step 1. Multiply the **mass of the given** by a conversion factor formed from the molar mass of the given.

Step 2. Multiply the number of **moles of the given** by a conversion factor composed of the coefficients in the balanced equation of the given and unknown substances.

Step 3. Multiply the number of **moles of the unknown** by the molar mass of the unknown.

Sample Problem. How many grams of oxygen will react with 24.0 grams of hydrogen to form water? Hydrogen is the given. Oxygen is the unknown.

Given: 24.0 g H_2 _Unknown:_ ? g O_2

Step 1: _Convert from "mass of the given" to "moles of the given."_ Step 1 answers the question, "How many moles of H_2 are there in 24.0 g H_2?" In shorthand: ? mol H_2 = 24.0 g H_2. The answer is found by dividing the mass of the given by the molar mass of the given. The molar mass of hydrogen is 2.02 grams per mole. The relationship between grams of H_2 and one mole of H_2 can be written in two ways, as follows:

$\dfrac{2.02 \text{ g } H_2}{1 \text{ mole } H_2}$ This expression is the molar mass. It says that there are 2.02 grams per mole of hydrogen.

$\dfrac{1 \text{ mole } H_2}{2.02 \text{ g } H_2}$ This expression is the reciprocal of molar mass. The reciprocal is the fraction turned upside down. Multiplying by the reciprocal of the molar mass is the same as dividing by the molar mass. In Step 1, you multiply the given mass by the reciprocal of the molar mass. This reciprocal is called conversion factor 1 in Fig. 28-2.

(Step 1) $\quad$? mol H_2 = 24.0 g H_2 $\times$ $\dfrac{1 \text{ mol } H_2}{2.02 \text{ g } H_2}$ = 11.9 mol H_2

given	**conversion**	**moles of**
mass	**factor**	**the given**
(Box 1)	**(Factor 1)**	**(Box 2)**

Look at the units in Step 1. The unit g H_2 appears twice, and divides out. That leaves the unit mol H_2. This is the unit the answer should have. Using the conversion factor makes this possible.

Step 2: *Convert from moles of given to moles of unknown.* Step 2 answers the question " How many moles of the unknown will react with the moles of the given found in Step 1?"

In Step 2, moles of the given must be multiplied by factor 2. Factor 2 is a fraction formed from coefficients in the balanced equation. The balanced equation is $2H_2 + O_2 \rightarrow 2H_2O$. The coefficient for the given is 2 ($2H_2$). The coefficient for the unknown is 1 (O_2). Two factors can be formed from these coefficients.

$$\frac{1 \text{ mole } O_2}{2 \text{ mole } H_2} \qquad\qquad \frac{2 \text{ mole } H_2}{1 \text{ mole } O_2}$$

The first factor provides the correct unit for the answer, since you want to divide out mol H_2.

(Step 2) $\quad$? mol O_2 = 11.9 mol H_2 $\times$ $\dfrac{1 \text{ mol } O_2}{2 \text{ mol } H_2}$ = 5.95 mol O_2

moles of	**conversion**	**moles of**
the given	**factor**	**the unknown**
(Box 2)	**(Factor 2)**	**(Box 3)**

Step 3: *Convert from moles of unknown to mass of unknown.* Step 3 answers the question "How many grams of the unknown are there in the number of moles found in Step 2?" Restate Step 3 using shorthand notation: ? g O_2 = 5.95 mol O_2. In Step 3, moles of the unknown must be multiplied by factor 3. The factor is formed from the molar mass of the unknown. The molar mass of oxygen is 32.0 g.

(Step 3) $\quad$? g O_2 = 5.95 mol O_2 $\times$ $\dfrac{32.0 \text{ g } O_2}{1 \text{ mol } O_2}$ = 190 g O_2

moles of	**molar mass**	**mass of**
unknown	**of unknown**	**the unknown**
(Box 3)	**(Factor 3)**	**(Box 4)**

The three steps are finished. The final answer: 190 g O_2.

You can combine Steps 1, 2, and 3 of a mass-mass problem as shown in Fig. 28-3.

Fig. 28-3

STEPS 1, 2, AND 3 COMBINED INTO ONE STEP

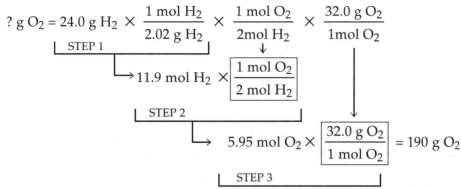

$$? \text{ g O}_2 = 24.0 \text{ g H}_2 \times \frac{1 \text{ mol H}_2}{2.02 \text{ g H}_2} \times \frac{1 \text{ mol O}_2}{2\text{mol H}_2} \times \frac{32.0 \text{ g O}_2}{1\text{mol O}_2}$$

STEP 1

$$11.9 \text{ mol H}_2 \times \boxed{\frac{1 \text{ mol O}_2}{2 \text{ mol H}_2}}$$

STEP 2

$$5.95 \text{ mol O}_2 \times \boxed{\frac{32.0 \text{ g O}_2}{1 \text{ mol O}_2}} = 190 \text{ g O}_2$$

STEP 3

Check Your Understanding

Write the missing terms in the blanks.

3. The substance whose mass is given in the statement of the problem is called _____.

4. A fraction that has been turned upside down is the _____ of the original fraction.

5. A(n) _____ problem is one in which the mass of one substance in a reaction is given and you are asked to find the mass of the other.

What Do You Know?

Questions 6 to 10 refer to this reaction: $2\text{Na} + \text{Cl}_2 \rightarrow 2\text{NaCl}$. Show your work for all problems on a separate sheet of paper.

6. If 24.0 grams of Cl_2 react, how many moles of Cl_2 will react? _____

7. How many moles of Cl_2 will react with 2.15 moles of Na? _____

8. If 5.76 moles of Na react, how many moles of NaCl will be produced? Assume there is enough Cl_2 to use up all the Na. _____

9. A reaction produces 4.73 moles of NaCl. How many grams of NaCl are produced? _____

10. During a reaction, 18.4 grams of Cl_2 react. How many moles of NaCl will be produced? Assume there is enough Na to use up all the Cl_2.

_____

Stoichiometry II

Key Words

volume-volume problem:	problem in which volume of one substance in a reaction is given and volume of another is found
Avogadro's hypothesis:	equal volumes of all gases at the same temperature and pressure contain the same number of molecules
mass-volume problem:	problem in which mass of one substance in a reaction is given and volume of another is found; or volume of one is given and mass of another is found
molar volume:	volume that one mole of any gas occupies at STP; its value is 22.4 liters

KEY IDEAS

To solve a volume-volume problem, the unknown is multiplied by only one factor to find the answer. Mass-volume problems are three-step problems. Three factors are used to obtain an answer. Both types of problems make use of molar volume.

When pharmaceutical chemists make a new medication, they must know how much of each reactant to use. This knowledge comes from an understanding of stoichiometry.

Fig 29-1

VOLUME-VOLUME
PROBLEM

Volume
of the
given
FACTOR

⟶
One step

Multiply the volume
of the given by a
volume factor. The
result is the answer.

Volume-Volume Problems. In a **volume-volume problem**, both the given and the unknown are gases. The volume of one substance is given and the volume of another is found. Fig. 29-1 shows the general method used to solve a volume-volume problem.

Sample Problem 1. How many liters of ammonia will be produced when 1.56 liters of nitrogen react with excess hydrogen? The balanced equation for the reaction is

$$3H_2(g) + N_2(g) \longrightarrow 2NH_3(g)$$

Given: $1.56\ l\ N_2$ *Unknown:* $?\ l\ NH_3$

Solution. In all chemical equations, coefficients tell the relative number of moles taking part in a reaction. In the equation above, the coefficient of N_2 is 1. The coefficient of NH_3 is 2. Therefore, the number of moles of NH_3 produced is two times the number of moles of N_2 reacting.

Avogadro's hypothesis (AH-voh-GAHD-rohs hy-PAH-thuh-sihs) provides a link between the number of molecules, number of moles, and the volumes of gases. It states that equal volumes of all gases at the same temperature and pressure contain the same number of molecules. This means that a mole of one gas will occupy the same volume as a mole of another gas. Two moles of a gas will occupy twice the volume of one mole of a gas.

In reactions between gases, Avogadro's hypothesis gives you a second meaning for coefficients. They tell the relative volumes of gases. The given is 1.56 l N_2. The coefficients tell you that twice that volume of NH_3 will be produced. Therefore, 3.12 l NH_3 will be produced. This is true if N_2 and NH_3 are at the same temperature and pressure.

Expression 1 is a setup for the problem. The quantity 1.56 l is multiplied by the volume factor. This factor comes from the coefficients 2 and 1 for NH_3 and N_2.

Expression 1 $? \, l \, NH_3 \; = \; 1.56 \, l \; N_2 \; \times \; \dfrac{2 \text{ vol. } NH_3}{1 \text{ vol. } N_2} \; = \; 3.12 \, l \, NH_3$

**given
volume**   **volume
factor**   **answer**

 1. Given the reaction $2CO(g) + O_2(g) \rightarrow 2CO_2(g)$, what volume of CO_2 will be produced when 8.45 l of CO reacts with enough oxygen? Show all work. _____

_____

Mass-Volume Problems. In a **mass-volume problem**, the given is either a mass or a volume. If the given is a mass, then the unknown is a volume. If the given is a volume, the unknown is a mass.

Mass-volume problems are three-step problems. First consider problems in which the given is a mass. See Fig. 29-2 for a diagram of the general method. The problem that follows is an example.

Fig 29-2

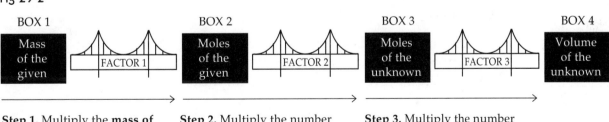

Step 1. Multiply the **mass of the given** by a factor formed from the molar mass of the given.

Step 2. Multiply the number of **moles of the given** by a factor composed of the coefficients in the balanced equation of the given and unknown substances.

Step 3. Multiply the number of **moles of the unknown** by the molar *volume* of the unknown.

Sample Problem 2. How many liters of oxygen at STP will react with 24.0 g H_2 to form water? Hydrogen is the given. Oxygen is the unknown.

Step 1: *Convert from "mass of the given" to "moles of the given."* This problem and the sample problem in the previous lesson are quite similar. In Lesson 28, the given was 24.0 g H_2. In this problem, it is the same mass. In Lesson 28, the unknown was the mass of O_2 reacting. In this problem, it is the volume of O_2 reacting. Steps 1 and 2 of both problems are identical.

(Step 1) $? \text{ mol } H_2 = 24.0 \text{ g } H_2 \times \dfrac{1 \text{ mol } H_2}{2.02 \text{ g } H_2} = 11.9 \text{ mol } H_2$

given	**conversion**	**moles of**
mass	**factor**	**the given**
(Box 1)	**(Factor 1)**	**(Box 2)**

Step 2: *Convert from moles of given to moles of unknown.*

(Step 2) $? \text{ mol } O_2 = 11.9 \text{ mol } H_2 \times \dfrac{1 \text{ mol } O_2}{2 \text{ mol } H_2} = 5.95 \text{ mol } O_2$

moles of	**conversion**	**moles of**
the given	**factor**	**unknown**
(Box 2)	**(Factor 2)**	**(Box 3)**

Step 3: *Convert from "moles of the unknown" to "volume of the unknown."* Step 3 answers the question "How many liters of the unknown are there in the number of moles found in Step 2?" This can be restated using shorthand notation as "$? \, l \, O_2 = 5.95 \text{ mol } O_2$." In Step 3, moles of the unknown must be multiplied by factor 3.

The link between moles and volume is the molar volume. The **molar volume** is the volume that one mole of any gas occupies at STP. The molar volume is 22.4 *l* at STP. Factor 3 in mass volume problems is formed from the molar volume. There are two factors. One is the reciprocal of the other.

$$\dfrac{22.4 \, l \, O_2}{1 \text{ mol } O_2} \qquad\qquad \dfrac{1 \text{ mole } O_2}{22.4 \, l \, O_2}$$

The first factor gives the correct units for the answer.

(Step 3) $? \, l \, O_2 = 5.95 \text{ mol } O_2 \times \dfrac{22.4 \, l \, O_2}{1 \text{ mol } O_2} = 133 \, l \, O_2$

moles of	**Conversion**	**Volume of the**
the unknown	**factor**	**unknown**
(Box 3)	**(Factor 3)**	**(Box 4)**

That ends the three steps. The final answer is: 133 l O_2. This is the correct volume at STP.

Mass-Volume Problems—Volume Given. In Sample Problem 2, a mass was the given. Sometimes the volume is the given. These problems are similar. However, all three factors are the reciprocals of those used when the given is a mass.

All three steps of Sample Problem 2 can be combined, as shown in Fig. 29-3.

Fig. 29-3

STEPS 1, 2, AND 3 COMBINED INTO ONE STEP

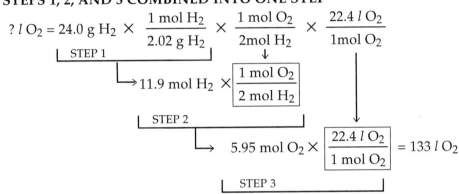

$$? \, l \, O_2 = 24.0 \text{ g } H_2 \times \frac{1 \text{ mol } H_2}{2.02 \text{ g } H_2} \times \frac{1 \text{ mol } O_2}{2 \text{ mol } H_2} \times \frac{22.4 \, l \, O_2}{1 \text{ mol } O_2}$$

STEP 1

$$\rightarrow 11.9 \text{ mol } H_2 \times \boxed{\frac{1 \text{ mol } O_2}{2 \text{ mol } H_2}}$$

STEP 2

$$\rightarrow 5.95 \text{ mol } O_2 \times \boxed{\frac{22.4 \, l \, O_2}{1 \text{ mol } O_2}} = 133 \, l \, O_2$$

STEP 3

Check Your Understanding

Write the missing term in each blank.

2. The _____ is the volume occupied by one mole of a gas at STP.

3. The numerical value of the molar volume is_____ at STP.

Questions 4 to 7 refer to the reaction $2CO + O_2 \rightarrow 2CO_2$. Show all work for these problems. (Both gas volumes are measured at the same temperature and pressure.)

4. What volume of O_2 gas will react with 18.0 l CO?

_____

5. What volume of CO gas will react with 8.76 l O_2? _____

6. What volume of CO_2 gas at STP will be produced when 6.88 g O_2 react with excess CO? _____

7. What mass of CO_2 gas will be formed from 1.24 l O_2 at STP? _____

_____

Summary

- The formula mass of a compound is the sum of the atomic masses of all the atoms indicated by the formula of the compound. The gram formula mass is numerically the same as the formula mass, but it uses grams as a unit.

- There are 6.02×10^{23} particles, usually atoms or molecules, in one gram formula mass. This number of particles is called a mole. It's also called Avogadro's number.

- The mass of one mole of atoms, molecules, or formula units is called the molar mass.

- Chemical formulas contain quantitative information. An empirical formula has the smallest possible subscripts. It tells the relative number of each kind of atom in the compound.

- A molecular formula tells the number of atoms of each element in a molecule. A molecular formula is sometimes an empirical formula.

- The known masses of the elements making up a sample of a compound can be used to calculate the percentage composition of a compound.

- The empirical formula of a compound can be determined from percentage composition. If the percentage composition and formula mass of a compound are known, its molecular formula can be determined.

- Chemical equations are commonly classified into four groups. These groups are synthesis, decomposition, single replacement, and double replacement.

- Balanced equations show that the number of atoms in the reactants and products are conserved during a chemical reaction. Balanced equations are needed to solve mass-mass, mass-volume, and volume-volume problems.

- Coefficients are numbers placed before formulas in a chemical equation. A coefficient indicates the number of formula units represented by the formula.

- One mole of every gas occupies the same volume at STP. That volume is 22.4 l. This volume is called the molar volume.

- The molar volume is used as one of the factors in mass-volume problems.

For Your Portfolio

1. Work in groups of two. The first student is to describe to the other how to find the formula mass of the following: **(a)** SO_2 **(b)** $Ca(OH)_2$ **(c)** $(NH_4)_2S$. The second student is to explain to the first the relationship between mole, gram formula mass, and molar mass.

2. Working in groups of six, students are to demonstrate the reaction between hydrogen and oxygen to produce water. Four students are to be hydrogen atoms. Two are to be oxygen atoms. Work together to show the arrangement of the atoms both before and after the reaction occurs.

3. Two carbon monoxide—CO—molecules react with an oxygen molecule to produce two carbon dioxide molecules—CO_2. Use black circles for the carbon atoms. Use white circles for oxygen atoms. On the left side of a sheet of paper, show the arrangement of the atoms before the reaction. On the right side, show their arrangement after the reaction. Use Fig. 26-2 on page 125 as an aid in making your drawing. Then for the equation, prepare a chart like the one shown in Fig. 26-3 on page 127.

4. Work in groups of four to represent atoms of different elements. One student is to be a magnesium atom. Another is to be a nickel atom. The remaining two are to be chlorine atoms. Show what happens during a single replacement reaction in which magnesium replaces nickel from the compound nickel chloride—$NiCl_2$.

5. A solution of barium nitrate—$Ba(NO_3)_2$— reacts with a solution of sodium sulfate —Na_2SO_4— to produce a precipitate of barium sulfate—$BaSO_4$. Prepare a diagram for this reaction similar to the one shown in Fig. 27-5 on page 130.

6. Eight grams of oxygen react with excess hydrogen to produce water. Draw molecular models to show this reaction. Write a balanced equation for the reaction. Tell how many grams of water will be produced.

In the blank spaces in Column I, write the letter of the item in Column II that best matches it. Oxygen has an atomic mass of 16.0.

	Column I		Column II
_____	1. formula mass of O_2	A.	32.0 grams
_____	2. number of atoms in one formula unit of O_2	B.	6.02×10^{23} O_2
_____	3. number of molecules in one mole of O_2	C.	32.0
_____	4. molar mass of O_2	D.	12.04×10^{23}
_____	5. number of atoms in one mole of O_2	E.	2

Fill in the blanks with the letter of the best choice. (A periodic table may be used in taking this part of the test.)

_____ 6. The formula mass of Na_2CO_3 is

    (a) 51.   **(b)** 74.   **(c)** 83.   **(d)** 106.

_____ 7. Which of the following is an empirical formula?

    (a) Fe_2O_3 **(b)** H_2O_2 **(c)** C_2H_4 **(d)** $C_4H_8O_4$

_____ 8. The empirical formula of benzene is CH. Its formula mass is 78. What is its molecular formula?

    (a) CH   **(b)** C_2H_4 **(c)** C_3H_3 **(d)** C_6H_6

_____ 9. Which coefficients balance the following equation?

$$C_2H_6 + O_2 \longrightarrow H_2O + CO_2$$

    (a) 1,1,1,1   **(b)** 2,8,6,4   **(c)** 2,7,6,4   **(d)** 1,7,6,3

Answer one of following questions.

10. **a.** What does "one mole of oxygen—O_2" mean? What is the relationship between it and the gram formula mass of O_2?

    _____

    _____

    b. What is the relationship between one mole of oxygen—O_2—and 22.4 l of oxygen at STP? Explain in detail how you arrived at your answer.

    _____

    _____

Solutions

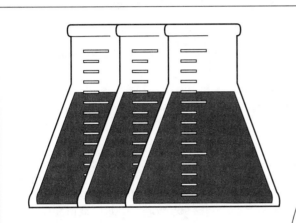

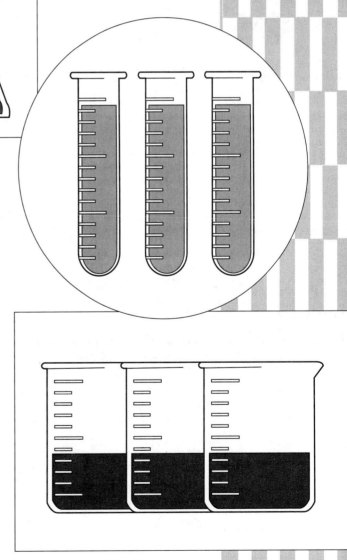

Do you ever swim in a pool? If so, you may know that the water in the pool must be tested to see if its pH is correct. This test helps the pool manager keep the chemicals in the water at the right level to prevent the spread of disease-causing germs. Do you have a fish tank? If so, you may test the water in your tank to see if its pH is correct. Did you know that environmental technicians also check the pH in lakes, ponds, and rivers? You guessed it—their concern is for the health of the fish and other wildlife living in these bodies of water.

Water in a pool, a fish tank, the ocean, a river, a lake, or even in your home is not pure. Even if it looks clear, water is almost always a mixture of materials. Water becomes a mixture as substances dissolve in the water. In this unit, you will find out what pH is and more about solutions.

Introduction to Solutions

solution:	mixture of substances that are evenly spread throughout each other; particles in a solution are molecular or ionic in size
solvent:	part of a solution that is usually present in the larger amount
solute:	part of a solution that is usually present in the smaller amount
dissolves:	breaks up into the smallest particles of that substance so that a solution forms
electrolyte:	substance that will conduct electricity and whose water solution contains ions

KEY IDEAS

A solution is a mixture of substances: the solvent and one or more solutes. The particles of solute are molecular or ionic in size and are spread evenly throughout the solvent. When water is the solvent and an ionic compound is the solute, the ions separate in the solution. Three types of solutions are gas solutions, liquid solutions, and solid solutions.

Since many materials mixed with the ocean's water are dissolved, the ocean can be thought of as a large solution. Technicians working with the living and nonliving natural resources of the ocean need to know how solutions affect these resources.

Parts of a Solution. A **solution** (suh-LOO-shuhn) is a mixture of two or more substances that are spread evenly throughout the mixture. The **solvent** (SAHL-vuhnt) is the part of the solution that is usually present in the larger amount. The **solute** (SAHL-yoot) is the part of the solution that is usually present in the smaller amount.

When a substance **dissolves** (dih-ZAHLVS), it breaks up into the smallest particles of that substance. For a molecular compound, these particles are molecules. For an ionic compound, these particles are ions. The dissolved substance is the solute of a solution.

A solution can have one or more solutes. For example, you can make a solution in which water is the solvent and sugar, broken down into molecules, is the solute. You can also make a solution in which water is the solvent and salt, broken down into ions, is the solute. In another solution, water can be the solvent and both salt and sugar can be solutes.

Properties of Solutions. Solutions share certain properties. These properties include the following:

- The particles of solvent and solute in solutions are spread evenly throughout the mixture.

- The particles of the solute are molecular or ionic in size.

- A solid solute can usually be separated from a liquid solvent by physical means, such as evaporation.

- The amounts of solute and solvent in a solution may vary within limits.

Molecules and Ions in Solutions. The action of a solvent on a solute forms a solution. Fig. 30-1 shows sugar dissolving in water. Water is the solvent and sugar is the solute. Molecules of water gather around the sugar crystal. There is a force of attraction between the solvent (water) and the solute (sugar). This force causes sugar molecules to leave the crystal and become dissolved in the water.

 1. In the solution shown in Fig. 30-1, sugar is the solute because _____

_____.

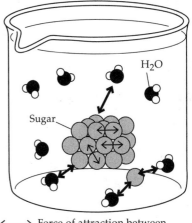

Fig. 30-1

⟷ Force of attraction between sugar molecules.

⟷ Force of attraction between sugar molecules and water molecules.

When an ionic compound is the solute and water is the solvent, the ions separate. Fig. 30-2 shows sodium chloride—NaCl—dissolving. Sodium chloride is ionically bonded. It consists of sodium ions—Na^+—and chloride ions—Cl^-. When sodium chloride dissolves, its Na^+ ions and Cl^- ions separate.

 2. In the solution shown in Fig. 30-2, is sodium chloride the solute or the solvent? _____

_____

Electrolytes. Substances whose water solutions contain ions will conduct electricity. Such substances are called **electrolytes** (ee-LEHK-troh-lyets). Sodium chloride is an electrolyte. Covalent compounds usually do not ionize when they dissolve in water. These solutions do not conduct electricity and are called nonelectrolytes. Sugar, for example, is a covalent compound that is a nonelectrolyte. Acids, however, are covalent compounds that do ionize when they dissolve in water. So an acid, such as hydrochloric acid, is an electrolyte.

Fig. 30-3 on page 146 shows a laboratory method used to test electrolytes. A light bulb is connected in a circuit with a low-voltage power source or battery. If the solution contains an electrolyte, the bulb will glow.

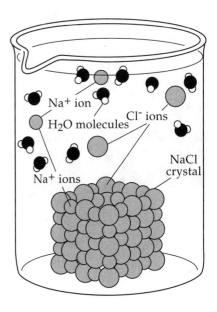

Fig. 30-2

Na^+ ion

H_2O molecules

Cl^- ions

Na^+ ions

NaCl crystal

Fig. 30-3   Laboratory method to test electrolytes

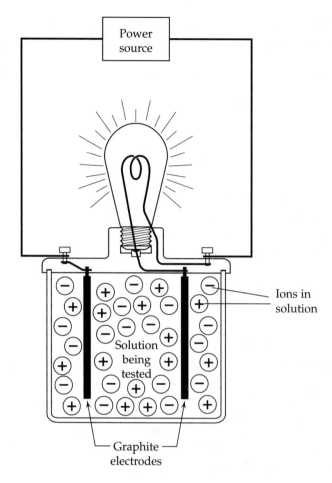

 3.   **Which substance, sugar or hydrochloric acid, is an electrolyte?**

_____

Solutions can be classified as gas solutions, liquid solutions, and solid solutions. The table describes and gives examples of each type of solution.

Type	Description of solution	Examples
Gas	Both solvent and solute are gases or vapors.	Air, made up mostly of nitrogen and oxygen; other gases
Liquid	Liquid solvent in which the solute is (a) a gas, (b) a liquid, or (c) a solid	(a) Soda water, made of solvent: water solute: carbon dioxide (b) Antifreeze, made of solvent: water solute: ethylene glycol (c) Vinegar, made of solvent: water solute: acetic acid
Solid	Both solvent and solute are solids.	Alloys of metals, such as brass, made of copper and zinc

Use the terms *electrolyte, solute, solution, solvent* to fill in the blanks. A term may be used more than once.

4. A mixture in which the particles are the size of molecules is a(n)

   _____.

5. The _____ of a solution is present in the larger amount.

6. The _____ of a solution is present in the smaller amount.

7. A substance whose water solution conducts electricity is a(n)

   _____.

8. In soda water, carbon dioxide is the _____ and water is

   the _____.

If the statement is correct write the word *True*. If the statement is incorrect write the word *False*.

9. _____ The smallest particle of a solute in a solution can be a molecule or an ion.

10. _____ A solute must be separated from its solvent by chemical means.

11. _____ A solution can only be made from the same amounts of solute and solvent.

12. _____ Hydrochloric acid is an electrolyte.

Underline the term in parentheses that correctly completes each statement.

13. When salt dissolves in water, it breaks up into Na^+ and Cl^- (molecules, ions).

14. When sugar dissolves in water, the solute is the (water, sugar).

15. Antifreeze is a solution in which a (gas, liquid) is the solute.

16. Brass is a solution in which a (liquid, solid) is the solvent.

Lesson 31 Solubility

KEY IDEAS

Solutions exist in different strengths. There are different ways of showing the amount of solute in a solution. These ways include solubility—the concentration in grams of solute per 100 grams of solvent; molarity—moles per liter of solution; and molality—moles per kilogram of solvent.

Carbonated soft drinks are among the many solutions you use. Technicians in the carbonated beverage industry need to know how to dissolve exact amounts of sweeteners, flavors, and carbon dioxide in water.

Solubility. A soluble substance is able to dissolve in a solvent. The amount of solute contained in a given amount of a solvent is the **concentration** (kahn-sun-TRAY-shuhn) of a solution. **Solubility** (sahl-yoo-BIHL-uh-tee) is the amount of a solute that will dissolve in a definite amount of solvent at a given temperature. Solubility can be expressed in terms of comparisons. Some materials are soluble; others are slightly soluble.

The chart in Fig. 31-1 shows solubilities of some substances in water. For example, the chart shows that barium chloride is soluble, while barium sulfate is insoluble, or not soluble.

 1. Will calcium nitrate dissolve in water? _____

 2. What is the solubility of silver iodide? _____

Fig. 31-1

SOLUBILITIES IN WATER										
d — decomposes i — insoluble s — soluble ss — slightly soluble x — not isolated	acetate	bromide	carbonate	chloride	chromate	hydroxide	iodidce	nitrate	phosphate	sulfate
Ammonium	s	s	s	s	s	s	s	s	s	s
Barium	s	s	i	s	i	s	s	s	i	i
Calcium	s	s	i	s	s	ss	s	s	i	ss
Copper II	s	s	i	s	i	i	x	s	i	s
Iron II	s	s	i	s	x	i	s	s	i	s
Lead	s	ss	i	ss	i	i	ss	s	i	i
Magnesium	s	s	i	s	s	i	s	s	i	s
Potassium	s	s	s	s	s	s	s	s	s	s
Silver	ss	i	i	i	ss	x	i	s	i	ss
Sodium	s	s	s	s	s	s	s	s	s	s

Temperature and Solubility. Temperature affects solubility. Study the graph in Fig. 31-2. For example, the graph shows that 30 grams of KCl will dissolve in 100 grams of water at 10°C. At 60°C, 45 grams of KCl will dissolve in 100 grams of water.

 3. How much $NaNO_3$ will dissolve in 100 g of water at 30° C? _____

There is a limit to solubility at any temperature. For example, no more than 22 grams of KNO_3 can dissolve in 100 grams of water at 10° C. At this point, the KNO_3 solution is saturated. A **saturated solution** (SACH-uh-rayt-ihd) contains the greatest amount of solute that can be dissolved in a solvent at a given temperature. As the temperature increases, the solubility of most solids increases.

 4. How much NaCl will saturate 100 g of water at 50°C? _____

Unlike solids, the solubility of gases in water decreases as temperature increases. Both NH_3 and SO_2 are gases. Note in the solubility curves that the solubility of these gases decreases as the temperature increases.

 5. How much does the solubility of NH_3 decrease between 10°C and 90°C? _____

Molarity. The concentration of a solution in moles per liter is **molarity** (moh-LAIR-uh-tee). A one-molar solution contains one mole of solute dissolved in a liter of solution. The math equation for molarity is:

$$\text{molarity } (M) = \text{moles of solute} \div \text{liters of solution}$$

Fig. 31-2

SOLUBILITY CURVES

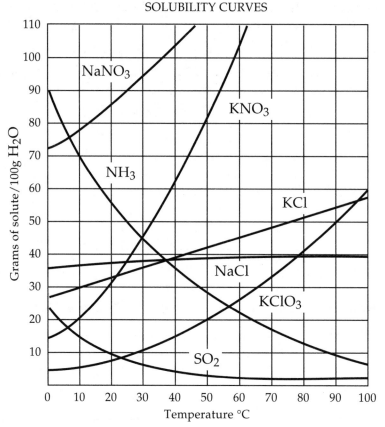

Sample Problem 1: What is the molarity of a sodium hydroxide (NaOH) solution containing 4.00 moles of solute in 2.00 liters of solution?

$$4.00 \text{ moles of solute} \div 2.00 \text{ liters of solution} = 2 \text{ mol/liter}$$
$$M = 2.00$$

Sample Problem 2: What is the molarity of a solution that contains 112 grams of KOH in 4.00 liters of solution? First, find the number of moles of KOH in 112 grams of KOH.

$$\text{mol} = \text{mass} \div \text{gram formula mass (GFM)}$$
$$\text{mol} = 112\text{g KOH} \div 56.0 \text{ grams per mole}$$
$$\text{mol} = 2.00$$

Next, substitute the number of moles and liters into the equation:

$$\text{molarity } (M) = \text{moles of solute/liters of solution}$$
$$M = 2.00 \text{ mol/4.00 liters of solution}$$
$$M = 0.500$$

The next sample problem shows how to find the amount of solute in solution, using the two equations from Sample Problems 1 and 2.

Sample Problem 3: How many grams of $NaNO_3$ are dissolved in 0.50 liters of 1.5 M solution? First, find the number of moles of $NaNO_3$ in the solution. Substitute the given values for molarity and volume into the equation:

$$M = mol/l$$
$$1.5 = mol/0.50 \; l$$
$$mol = 0.75$$

Next, change moles to grams using the equation:

$$mol = mass/GFM$$
$$0.75 = mass/85 \text{ g per mol}$$
$$mass = 63.75 \text{ g}$$

 6. How many grams of KCl are needed to prepare 2.0 liters of 2.0 M solution? _____

_____

It is often necessary to add more solvent to, or dilute, a solution of known molarity. Study the next sample problem.

Sample Problem 4: How much water is needed to make 20 ml of 6.0 M HCl from 12 M HCl? To solve the problem use this equation:

$$V_1M_1 = V_2M_2$$

In this equation, V_1 = the starting volume
M_1 = the starting molarity
M_2 = the final molarity
V_2 = the final volume

Solving the above problem: $V_1 (12 \; M) = (20 \text{ ml})(6M)$
$$V_1 = 10 \text{ m}l \text{ HCl}$$

The amount of water needed is found by subtracting the starting volume from the final volume.

$$20 \text{ m}l - 10 \text{ m}l = 10 \text{ m}l \text{ of water}$$

Fig. 31-3 shows 10 ml of 12 M HCl and 10 ml of water used to make 20 ml of 6.0 M solution.

Fig. 31-3

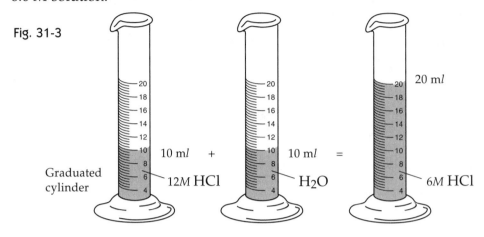

Molality is another way to express the concentration of a solution. **Molality (m)** (moh-LAL-uh-tee) is the number of moles of solute per kilogram of solvent. The math equation for molality is

molality (m) = moles of solute/ kilograms of solvent

As you will see, molality is a useful tool to explain some topics in Lesson 32.

Fig. 31-4 shows how to prepare a 1 M solution in three steps. First, place one mole of solute into a one-liter volumetric flask. Next, add enough solvent to dissolve the solute. Then, add solvent to the one-liter mark and mix the solution completely.

Fig. 31-4

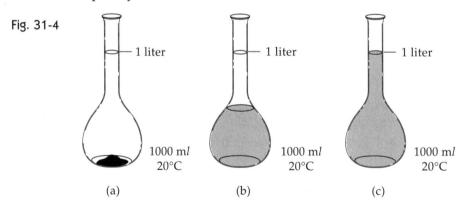

The data shown in the graph in Fig. 31-2 on page 150 can be stated in chart form. Study Fig. 31-5. It shows solubilities of several substances per 100 grams of water at different temperatures.

Fig. 31-5

	Grams of solute per 100 g water			
Temperature	NaCl	KCl	NH$_4$Cl	NH$_3$
30 °C	38 g	37 g	42 g	45 g
60 °C	39 g	45 g	56 g	23 g
90 °C	40 g	54 g	72 g	10 g

Use the terms *molarity, solubility, molality,* and *saturated* to fill in the blanks.

7. _____ is the concentration of a solution in moles per kilogram of solvent.

8. _____ is the amount of solute that will dissolve at a given temperature.

9. A _____ solution contains the greatest amount of solute that will dissolve at a given temperature.

10. _____ is the concentration of a solution in moles of solute per liter of solution.

Use the chart in Fig. 31-1 on page 149 and the graph in Fig. 31-2 on page 150 to answer the following.

11. How many grams of SO_2 are needed to saturate 100 g of water at 30°C? _____

12. At what temperature will 80 grams of KNO_3 saturate 100 g of water _____

13. Which bromide is insoluble? _____

14. All nitrate compounds are _____.

15. Name two calcium compounds that are slightly soluble. _____ and _____

What Do You Know?

Solve the following problems. You may use the periodic table on page 252.

16. What is the molarity of a solution that has 120 grams of NaOH in 500 m*l* of solution? _____

17. If 20 m*l* of 6.0 *M* HCl is diluted to a total volume of 60 m*l*, what is the molarity of the new solution? _____

18. How many grams of HCl are needed to make 200 m*l* of 1.50 *M* solution? _____

19. What volume of water is needed to make 310 m*l* of 3.00 *M* H_2SO_4 from 18 *M* H_2SO_4? _____

Lesson 32
Colligative Properties

Lesson

Key Words

colligative properties:	properties that depend upon the number of solute particles in solution and not the properties of the solute
molal boiling point elevation constant:	amount a one-molal solution of nonelectrolyte will raise the boiling point of a solution
molal freezing point depression constant:	amount a one-molal solution of nonelectrolyte will lower the freezing point of a solution
osmosis:	movement of solvent molecules from an area of high solvent concentration to an area of low solvent concentration
osmotic pressure:	pressure that must be applied to a solution to prevent osmosis

KEY IDEAS

Freezing point, boiling point, and osmotic pressure are all **colligative properties** (kuh-LIG-uh-tiv). These properties depend upon the number of solute particles in solution rather than the properties of the solute itself. The lowering of the freezing point of water is a colligative property that explains how antifreeze works in car engines. Osmotic pressure is a colligative property that needs to be understood by health-care workers who administer intravenous (IV) solutions.

Boiling Point Elevation. The boiling point of a solution depends on the concentration of solute particles. As the concentration of the solution increases, its boiling point also increases. For example, a one-molal (1 *m*) solution of sugar in water boils at 100.52°C. A two-molal (2 *m*) solution of sugar water boils at 101.04°C. The **molal boiling point elevation constant** (ehl-uh-vay-shuhn KAHN-stuhnt) for water is 0.52°C per molal solution.

 1. What is the boiling point of a 3 *m* sugar solution? _____

When most molecular compounds dissolve, they separate into molecules. These compounds are nonelectrolytes. Sugar is an example of a nonelectrolyte. When ionic compounds dissolve, they separate into ions. These compounds are electrolytes. NaCl is an example of an electrolyte. When it dissolves, each NaCl unit separates into two ions. Thus, one mole of NaCl will produce twice as many dissolved particles as one mole of sugar. As a result, the boiling point of a one-molal (1 *m*) solution of NaCl is elevated twice as much as 0.52°C.

t of 2 *m* NaCl? _____

eezing points of solutions also depend on
les. The **molal freezing point depression**
N-stuhnt) for water is 1.86°C per molal
in water freezes at –1.86°C. A 2 *m* solution

of a 3 *m* sugar solution? _____

e boiling point or freezing point of any

= K *m* n

ige from the normal boiling or freezing
ression *constant* is K. The molality of the
icles per formula unit is n. For example,
chloride. You can tell how many solute
f ions in its formula. There are one Ca^{2+}
al of three particles for the formula unit.

g point of a 1.5 *m* $CaCl_2$ solution? Since
3.

n

2°C/*m*)(1.5 *m*)(3)

= 2.34°C

Boiling point = 100°C + 2.34°C = 102.34°C

In this expression, the temperature change (2.34°C) is added to the normal
boiling point of water (100°C). The result is the boiling point of the solution.

Osmosis. Membranes are thin covering sheets of material, such as the cell
membranes of the cells of the human body and other living things. These
membranes allow some molecules, but not others, to flow through them.
The movement of solvent molecules, usually water, through a membrane is
osmosis (ahs-MOH-sihs). Osmosis takes place across cell membranes.

Fig. 32-1

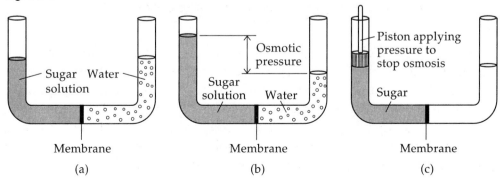

(a)   (b)   (c)

Part *a* of Fig. 32-1 shows a membrane that separates a sugar solution and water. The solvent—water—flows back and forth across the membrane. However, the rate of the flow of water is greater toward the solution with the higher concentration of sugar. This is because water molecules tend to move from an area of high water concentration to an area of low water concentration. As the flow of solvent continues, the liquid levels become uneven, as shown in Part *b* of the diagram. This difference in level causes a downward force known as **osmotic pressure** (ahs-MAHT-ihk PRESH-uhr).

If the same amounts of pressure were applied to the solute side of the tube, osmosis would stop. The number of solute particles determines osmotic pressure. A high-particle concentration results in a high osmotic pressure, as shown in Part *c* of Fig. 32-1.

Boiling points and freezing points of liquids are affected by the number of dissolved particles per formula unit. The chart below compares water solutions of some substances, their boiling points, and their freezing points.

Solute (*aq*)	Number of Particles per Formula Unit	Boiling point (°C)	Freezing point (°C)
KCl	2	101.04	-3.72
CaCl$_2$	3	101.56	-5.58
NaC$_2$H$_3$O$_2$	2	101.04	-3.72
Al(NO$_3$)$_3$	4	102.08	-7.44
CuSO$_4$	2	101.04	-3.72
C$_{12}$H$_{22}$O$_{11}$	1	100.52	-1.86

If you put a cucumber in a saltwater solution, osmosis takes place, as shown in Fig. 32-2. The cucumber loses water by osmosis. Water molecules flow in both directions across the cucumber's skin, but more water flows out of the cucumber than into it. This turns the cucumber into a pickle.

Fig. 32-2

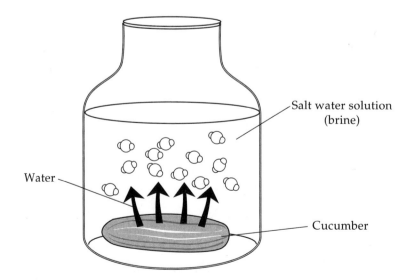

Salt water solution (brine)

Water

Cucumber

Fill in the blanks with the correct answers.

4. What is the boiling point of 2 *m* ZnSO₄ water solution? _____

5. What is the freezing point of 6 *m* C₁₂H₂₂O₁₁ water solution?_____

6. The movement of a solvent through a membrane is called _____.

7. The flow of solvent is always toward the _____ solution.

8. The force needed to stop the flow of molecules across a membrane is
   called _____.

Circle the letter of the correct answer.

9. What is the freezing point of a solution that contains one mole of
   MgBr₂ in 1000 g of water?

   a. 0.00°C   **b.** -1.86°C   **c.** -3.72°C.   **d.** -5.58°C

10. Which one-molal water solution will have the highest boiling point?

    a. NaCl   **b.** MgSO₄   **c.** CaI₂   **d.** C₆H₁₂O₆

11. As the molality of a solution increases, its osmotic pressure

    a. increases.   **b.** decreases.   **c.** remains the same.

12. As the molality of a solution increases, its freezing point

    a. increases.   **b.** decreases.   **c.** remains the same.

Answer the following in complete sentences.

13. Why is the boiling point elevation of a 1 *m* solution of NaCl twice as
    much as that of a 1 *m* solution of sugar? _____

    _____

    _____

14. Describe the flow of water molecules across a membrane between pure
    water and a solution of sugar. _____

    _____

    _____

Acids and Bases

indicator:	substance used to detect the presence of an acid or a base; acids and bases cause indicators to change color
Arrhenius acid:	substance that produces hydrogen ions when it is in water solution
Arrhenius base:	substance that produces hydroxide ions when it is in water solution
Bronsted-Lowry acid:	a proton donor
Bronsted-Lowry base:	a proton acceptor
hydronium ion:	a hydrated proton or H_3O^+

KEY IDEAS

Observing the properties of acids and bases has led to two main theories. One is the Arrhenius theory, which states that acids produce hydrogen ions and bases produce hydroxide ions. The other is the Bronsted-Lowry theory, which states that acids are proton donors and bases are proton acceptors.

In recent years, acid rain has become a serious environmental problem. Some technicians are working on ways to prevent acid rain from forming. Other workers are trying to cope with the effects of acid rain pollution that have already occurred.

Properties of Acids and Bases. Acids have the following observed properties:

- Acids dissolved in water are electrolytes, which conduct an electric current.

- Acids have a sour taste. Examples are the acids in vinegar and lemon juice.

- Acids react with many metals to produce hydrogen gas.

- Acids change the color of some indicators. An **indicator** (IN-duh-KAYT-uhr) is a substance used to detect the presence of an acid or a base. In the presence of an acid, blue litmus turns red, and red phenolphthalein becomes colorless.

- Acids neutralize bases to produce a salt and water.

Bases have the following observed properties:

- Bases dissolved in water are electrolytes.

- Bases feel slippery.

- Bases change the colors of some indicators. In the presence of a base, red litmus turns blue, and colorless phenolphthalein turns red.

- Bases neutralize acids to produce a salt and water.

 1. What color is litmus in the presence of an acid? _____

 2. What color is phenolphthalein in the presence of a base? _____

Arrhenius Theory. Arrhenius proposed a theory to explain the behavior of acids and bases. An **Arrhenius acid** (uh-RAY-nee-uhs) is a substance that produces hydrogen ions (H^+) as the only positive ions in water solution. Here is an example:

Equation 1 $HCl \ (in \ H_2O) \longrightarrow H^+ + Cl^-$

An **Arrhenius base** is a substance that produces hydroxide ions (OH^-) as the only negative ions in water solution. Here is an example:

Equation 2 $NaOH \ (in \ H_2O) \longrightarrow Na^+ + OH^-$

 3. Which symbol represents the hydrogen ion? The hydroxide ion?

_____

Bronsted-Lowry Theory. Bronsted and Lowry proposed another theory to explain acid and base reactions that take place in either a water or a nonwater medium. According to this theory, a **Bronsted-Lowry acid** (BRAHN-stehd LOW-ree) is a proton donor. A **Bronsted-Lowry base** is a proton acceptor.

Recall that the hydrogen atom consists of one proton and one electron. As shown in Fig. 33-1, when a hydrogen atom loses an electron, only a proton remains. Thus, a hydrogen ion is a proton.

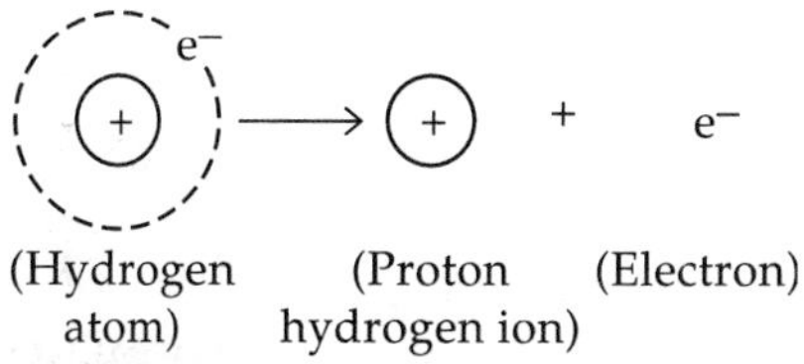

Equation 3 shows HCl reacting with H_2O to produce hydronium ion (H_3O^+).

Fig. 33-1

Equation 3 $HCl + H_2O \longrightarrow H_3O^+ + Cl^-$

(Hydrogen atom)   (Proton hydrogen ion)   (Electron)

Fig. 33-2

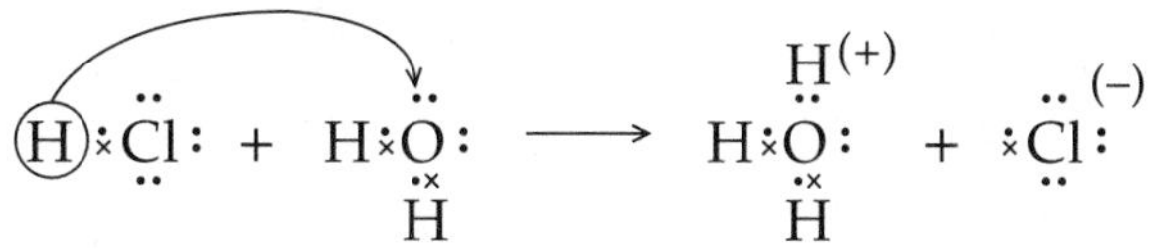

In the electron-dot diagrams shown in Fig. 33-2, you can see that a proton moves from the HCl to the H_2O. A hydronium ion—H_3O^+—is formed. The **hydronium ion** (hy-DROH-nee-uhm) is also called a hydrated proton because the proton is attached to a water molecule.

4. What is the formula for the hydronium ion? _____

An acid can give its proton to other substances besides water. In the reaction below, HCl loses its proton to ammonia, NH_3, forming an ammonium ion NH_4^+. This example shows that it is not necessary for the base to contain hydroxide OH^-. See Equation 4 and Fig. 33-3.

Equation 4 $$HCl + NH_3 \longrightarrow NH_4^+ + Cl^-$$

Fig. 33-3

A base, such as NaOH, accepts a proton from an acid, such as HCl.

Equation 5 $$NaOH + HCl \longrightarrow HOH + Na^+ + Cl^-$$

It is not necessary for a base to contain OH^-. For example, in Equation 4 in the reaction between HCl and NH_3, the base is NH_3.

TAKE ANOTHER LOOK

You've seen how electron-dot diagrams represent water molecules, hydroxide ions, and hydronium ions. Fig. 33-4 shows how these particles can be pictured as models made of spheres.

Fig. 33-4

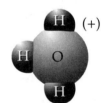

Water molecule
H_2O

Hydroxide ion
OH^-

Hydronium ion
H_3O^+

Indicators appear as different colors in acids and bases. The chart in Fig. 33-5 compares colors of various indicators.

Fig. 33-5

Table of Indicator Colors		
Indicator	**Color**	
	Acid	**Base**
alizarin yellow	yellow	violet
bromthymol blue	yellow	blue
litmus	red	blue
methyl red	red	yellow
phenolphthalein	colorless	red
phenol red	yellow	red

Fill in the blanks.

5. According to Arrhenius, an acid produces _____ ions and a base produces _____ ions.

6. The Bronsted-Lowry theory states that an acid is a(n) _____ and a base is a(n) _____.

7. A hydrated proton is called a(n) _____ ion and has the formula _____.

On the lines under the following equations, write the word *acid* or *base* to identify the substance as either a proton donor or a proton acceptor.

8. $HCl \quad + \quad H_2O \longrightarrow H_3O^+ \quad + \quad Cl^-$

   _____   _____

9. $H_2O \quad + \quad NH_3 \longrightarrow NH_4^+ + OH^-$

   _____   _____

Write the correct term in each blank.

10. One property of acids is their _____ taste.

11. In an acid solution, the color of litmus is _____.

12. Compounds that produce hydrogen ions in a water solution are _____.

13. In a base solution, the color of phenolphthalein is _____.

14. One property of bases is their _____ feel.

15. A(n) _____ is any substance used to detect the presence of an acid or a base.

If the statement is correct, write the word *True*. If the statement is incorrect, write the word *False*.

16. _____ Acids and bases are both electrolytes.

17. _____ A base is a proton acceptor.

18. _____ The formula for the hydronium ion is OH^-.

19. _____ A hydrated proton is called a hydroxide.

20. _____ An acid must always give its proton to water.

34 Describing Acid-Base Solutions

KEY IDEAS

In acid-base reactions, protons—H^+—move from one substance to another. Not all acids and bases lose or gain protons to the same degree. The extent of proton transfer determines acid or base strength. A pH number describes the concentration of hydrogen ions—H^+—or hydronium ions—H_3O^+.

Human blood is a slightly basic solution with a pH of about 7.4. Changes in the pH of the blood may occur when the body does not function properly. If the pH rises to near 8.0 or drops to below 6.8, the result can be fatal.

Proton Transfer. Recall that an acid is a proton donor. The proton, or H^+, is accepted by a base, which is a proton acceptor. For example, in a reaction between H_2SO_4 and H_2O, a proton moves from the H_2SO_4 to the H_2O, forming H_3O^+ and HSO_4^-.

Equation 1
$$\underset{\text{acid}}{H_2SO_4} + \underset{\text{base}}{H_2O} \longrightarrow H_3O^+ + HSO_4^-$$

The H_2SO_4 is the acid because it donates a proton—H^+. The H_2O is the base because it gains a proton—H^+. As a result of the H_2O gaining a proton, H_3O^+ is formed.

The reverse of this reaction can also occur. In this case, a proton moves from the H_3O^+ to the HSO_4^-, forming H_2O and H_2SO_4.

Equation 2
$$\underset{\text{acid}}{H_3O^+} + \underset{\text{base}}{HSO_4^-} \longrightarrow H_2SO_4 + H_2O$$

The H_3O^+ is an acid because it donates a proton—H^+—to the HSO_4^-. The HSO_4^- is the base because it gains a proton—H^+—from the H_3O^+.

When the products of a chemical reaction react to reform the reactants, the reaction is called a reversible reaction. Equation 3 combines Equation 1 with Equation 2 as one equation, showing a reversible reaction.

Equation 3 $\qquad$ $H_2SO_4 + H_2O \rightleftharpoons H_3O^+ + HSO_4^-$
$\qquad\qquad\qquad$ acid $\qquad$ base $\qquad\qquad$ acid $\qquad$ base

Acid-base Pairs. The H_2SO_4 became HSO_4^- when it lost, or donated, a proton. After an acid has donated a proton, the substance remaining is a base. This base forms a related pair with that acid. So H_2SO_4 and HSO_4^- are a **related acid-base pair.** The acid and base in this pair differ by only one proton.

The H_2O became H_3O^+ when it gained, or accepted, a proton. After a base has accepted a proton, the substance remaining is an acid. This acid forms a related pair with that base. So H_3O^+ and H_2O are a related acid-base pair. The acid and base in this pair differ by only one proton. Study Fig. 34-1, which shows acid-base pairs for the reaction in Equation 3.

Fig. 34-1 $\qquad\qquad$ $H_2SO_4 + H_2O \rightleftharpoons H_3O^+ \quad HSO_4^-$
$\qquad\qquad\qquad$ ACID $\quad$ BASE $\qquad\qquad$ ACID $\quad$ BASE

acid/base pairs

 1. On the lines under each substance in the equation, write the word *acid* or *base* to identify the substance.

$\qquad$ HCl $\quad + \quad$ H_2O $\rightleftharpoons$ H_3O^+ $\quad + \quad$ Cl^-

$\qquad$ (a) _____ $\quad$ (b) _____ $\qquad$ (c) _____ $\quad$ (d) _____

 2. The related acid-base pairs in Equation 2 are

$\qquad$ (a) _____ and (b) _____ .

Substances That Act As Acids or Bases. Some substances can act as either an acid or a base. When in the presence of an acid, such a substance acts as a base. When in the presence of a strong base, however, the same substance acts as an acid.

Water is an example of such a substance. When water donates a proton to NH_3, which is a strong base, the water is an acid.

Equation 4 $\qquad$ $H_2O + NH_3 \longrightarrow NH_4^+ + OH^-$
$\qquad\qquad\qquad$ acid $\quad$ base

When water accepts a proton from HCl, which is an acid, the water is a base.

Equation 5 $\qquad$ $HCl + H_2O \longrightarrow H_3O^+ + Cl^-$
$\qquad\qquad\qquad$ acid $\quad$ base

Water ionizes only slightly. When this happens, one water molecule donates a proton to another water molecule. Water, therefore, acts as both acid and base.

Equation 6 $$H_2O + H_2O \rightleftharpoons H_3O^+ + OH^-$$
$$\text{acid} \quad \text{base}$$

 3. **What is one substance with which water acts as an acid? What is a substance with which water acts as a base?**

(a) _____ and (b) _____.

Ionization Constants. An ionization constant, K_a, is used to compare the relative strengths of acids. To compute K_a for an acid, the concentration of the ions is divided by the concentration of the acid. A strong acid yields a large concentration of ions. A weak acid produces few ions in comparison to the number of acid molecules. So the K_a values of strong acids are larger than the K_a values of weak acids.

The chart in Fig. 34-2 lists some acids, the bases with which they form related pairs, and K_a values. The strong acids are at the top of the chart. Compare phosphoric acid—H_3PO_4—with acetic acid CH_3COOH. Phosphoric acid is the stronger acid and is higher on the chart. Also compare the K_a values of the two acids. The K_a of H_3PO_4 is 7.5×10^{-3}. This K_a is larger than the 1.8×10^{-5} value for CH_3COOH. A larger K_a means more ions and a stronger acid.

Fig. 34-2

Strengths of Acids		
Related acid-based pairs		
ACID	*BASE*	K_a
$HCl = H^+ + Cl^-$		large
$HNO_3 = H^+ + NO_3^-$		large
$H_2SO_4 = H^+ + HSO_4^-$		large
$HSO_4^- = H^+ + SO_4^{2-}$		1.2×10^{-2}
$H_3PO_4 = H^+ + H_2PO_4^-$		7.5×10^{-3}
$HNO_2 = H^+ + NO_2^-$		4.6×10^{-4}
$HF = H^+ + F^-$		3.5×10^{-4}
$CH_3COOH = H^+ + CH_3COO^-$		1.8×10^{-5}
$H_2CO_3 = H^+ + HCO_3^-$		4.3×10^{-7}
$HSO_3^- = H^+ + SO_3^{2-}$		1.1×10^{-7}
$H_2S = H^+ + HS^-$		9.5×10^{-8}
$H_2PO_4^- = H^+ + HPO_4^{2-}$		6.2×10^{-8}
$NH_4^+ = H^+ + NH_3$		5.7×10^{-10}
$HCO_3^- = H^+ + CO_3^{2-}$		5.6×10^{-11}
$HPO_4^{2-} = H^+ + PO_4^{3-}$		2.2×10^{-13}
$HS^- = H^+ + S^{2-}$		1.3×10^{-14}
$H_2O = H^+ + OH^-$		1.0×10^{-14}

4. Compare HF with H_2S. Which acid is stronger? _____.

Acidity as pH. The acidity of solutions can be stated in terms of **pH**. Neutral solutions have a pH value of 7. Acidic solutions have pH values less than 7. Basic solutions have values greater than 7.

$$pH < 7 \text{ Acidic Solution}$$
$$pH = 7 \text{ Neutral Solution}$$
$$pH > 7 \text{ Basic Solution}$$

Mathematically pH is the negative logarithm, to the base 10, of the concentration of the hydronium ion—H_3O^+. Brackets [ ] around a formula mean concentration in moles/liter.

Equation 7 $$pH = -\log [H_3O^+]$$

When water ionizes, hydronium—H_3O^+—and hydroxide—OH^-—ions are formed. K_w stands for the ionization constant of water. It has a value of 1.0×10^{-14}.

Equation 8 $$K_w = [H_3O^+] [OH^-] = 1.0 \times 10^{-14}$$

In pure water, $[H_3O^+] = [OH^-] = 1.0 \times 10^{-14}$. Therefore, the $[H_3O^+]$ and the $[OH^-]$ must both be 1×10^{-7} because $(1 \times 10^{-7})(1 \times 10^{-7}) = (1 \times 10^{-14})$.

Substituting the concentration of 1×10^{-7} into Equation 7, you can calculate the pH of water as 7.00.

The pH of a solution can be easily found with a calculator. Use your calculator and the following procedure shown in Fig. 34-3.

Fig. 34-3

1. Enter 1.0 Exp -7   **2.** Then push ($^1/_\times$)   **3.** Then push (Log)

Repeat the procedure to find the pH of a 0.1 M HCl solution. See Fig 34-4.

Fig. 34-4

1. Enter 0.1   **2.** Then push ($^1/_\times$)   **3.** Then push (Log)   Answer is pH = 1.00

You can estimate pH using the system shown in Fig 34-5.

Fig. 34-5 1.0×10^{-3}

If this number is exactly 1, then this number is the pH

5. (a) What is the pH of a solution with $[H_3O^+] = 1.0 \times 10^{-12}$? _____

(b) Is this solution an acid or a base? _____

TAKE ANOTHER LOOK box

The scale in Fig. 34-6 shows $[H_3O^+]$ and pH. On this scale, you can see that a solution with $[H_3O^+] = 1 \times 10^{-7}$ has a pH of 7 and is neutral. Acidic solutions have a pH less than 7. Basic solutions have a pH greater than 7.

Fig. 34-6

$[H_3O^+]$	1×10^0	1×10^{-1}	1×10^{-2}	1×10^{-3}	1×10^{-4}	1×10^{-5}	1×10^{-6}	1×10^{-7}	1×10^{-8}	1×10^{-9}	1×10^{-10}	1×10^{-11}	1×10^{-12}	1×10^{-13}	1×10^{-14}
pH	0	1	2	3	4	5	6	7 Neutral	8	9	10	11	12	13	14

Fig. 34-7 compares a strong acid and a weaker acid. The strong acid—HCl— produces many ions in solution. The weaker acid— HF—produces fewer ions.

Fig. 34-7

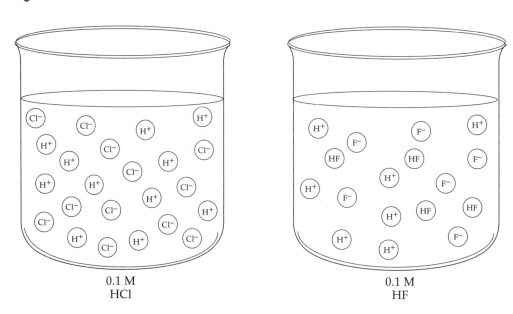

0.1 M
HCl

0.1 M
HF

Remember that pH is based upon the concentration of the hydronium (or hydrogen) ion. Low pH numbers mean a high hydronium (or hydrogen) ion concentration and a solution that is acidic. High pH numbers mean that many hydroxide ions are present and the solution is basic.

Use the key terms from the beginning of this lesson to fill in the blanks.

6. An acid and a base that differ by only one proton are called _____.

7. Relative strengths of acids are compared using _____.

8. The concentration of H_3O^+ in solution is expressed in terms of

_____.

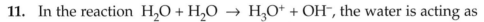

**What
Do You
Know?**

Circle the correct term.

9. The ionization constant, K_a, for acetic acid—CH_3COOH—is

    a. 1.2×10^{-2}.   **b.** 3.5×10^{-4}.   **c.** 1.8×10^{-5}.   **d.** 5.6×10^{-11}.

10. A sample of water contains

    a. equal concentrations of H_3O^+ and OH^-.
    b. greater concentrations of H_3O^+ than OH^-.

    c. lower concentrations of H_3O^+ than OH^-.
    d. no H_3O+ or OH^-.

11. In the reaction $H_2O + H_2O \rightarrow H_3O^+ + OH^-$, the water is acting as

    a. both an electron receiver and an electron donor.
    b. neither an electron receiver nor an electron donor.

    c. neither a proton donor nor a proton accepter.
    d. both a proton donor and a proton acceptor.

12. In the reaction $H_2S + H_2O \rightarrow H_3O^+ + HS^-$, a related acid base pair is

    a. H_2S and H_2O.
    b. H_2O and H_3O^+.

    c. H_3O^+ and HS^-.
    d. H_2O and HS^-.

13. What is the pH of a solution if the $[H_3O^+]$ is 1×10^{-8}?

    a. 1   **b.** 6   **c.** 8   **d.** 14

14. Pure water has a pH of

    a. 1×10^{-7}.   **b.** 7.   **c.** 1.   **d.** 1×10^{-14}.

Acid-Base Reactions

neutralization:	reaction between an acid and a base to make a salt and water
salt:	compound of a positive ion other than H^+ and a negative ion other than OH^-
titration:	process of finding the concentration of an unknown solution by reacting it with a standard solution
standard solution:	solution of known concentration
end point:	point in titration at which chemically equivalent amounts of acid and base are present
phenolphthalein:	indicator that is colorless in the presence of an acid and red in the presence of a base

KEY IDEAS

Acids and bases neutralize each other. If the strength of one of two solutions that neutralize each other is known, titration can be used to find the strength of the second solution.

Technicians in medical and in environmental laboratories use titration to analyze solutions. These workers need to be skilled in titration techniques and must be able to interpret the results.

Neutralization. The reaction that takes place when an acid and a base react to form a salt and water is **neutralization** (noo-truhl-ih-ZAY-shun). Look at these reactions:

Equation 1 $\quad HCl \quad + \quad NaOH \longrightarrow NaCl \quad + \quad H_2O$
hydrochloric   sodium   sodium   water
acid   hydroxide   chloride

Equation 2 $\quad HNO_3 \quad + \quad KOH \longrightarrow KNO_3 \quad + \quad H_2O$
nitric   potassium   potassium   water
acid   hydroxide   nitrate

In each reaction, an acid and a base neutralize each other. In each reaction, water and a **salt** are formed. A salt is a compound with a positive ion other than the hydrogen ion—H^+—and a negative ion other than the hydroxide ion—OH^-.

The salt NaCl was formed in the first reaction. The salt KNO_3 was formed in the second reaction. In each case, the salt is the result of the combination of the positive ion from the base with the negative ion of the acid.

 1. What salt would form if HCl neutralized KOH? _____

Titration. If the strength of only the acid or the base is known, the strength of the other solution can be measured by titration. **Titration** (ty-TRAY-shuhn) measures the concentration of an unknown solution by reacting it with a standard solution.

Fig. 35-1

To find the concentration of an acid, such as HCl, a burette such as the one in Fig. 35-1 is filled with a standard solution of a base, such as NaOH. The NaOH solution is titrated, or added in small amounts, into the HCl until the end point is reached. A **standard solution** is a solution of known concentration. The **end point** is that point in titration at which chemically equivalent amounts of acid and base are present.

The burette is a tool that can be used to measure the exact amount of a base of known concentration that will react with an acid of unknown strength. If phenolphthalein is present in the acid, a red color will appear at the end point. **Phenolphthalein** (fee-nohl-THAYL-een) is an indicator that is colorless in the presence of an acid and red in the presence of a base.

Sample Problem: Suppose 10 m*l* of a solution of HCl is titrated to the end point with 25.0 m*l* of 1.00 *M* NaOH. What is the molarity of the HCl solution?

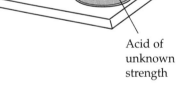

Burette

Clamp

Base of known concentration

Erlenmeyer flask

Acid of unknown strength

$$\text{molarity} = \text{moles/liters}$$
$$1.00\ M = \text{moles of NaOH}/0.0250 \text{ liters of NaOH}$$
$$\text{moles of NaOH} = 0.0250$$

The equation for the neutralization in this problem is

$$HCl + NaOH \longrightarrow NaCl + H_2O$$

In this equation, the moles of acid, HCl, neutralized is also 0.0250.

$$\text{molarity} = \text{moles/liters}$$
$$\text{molarity} = 0.0250 \text{ moles of HCl}/0.0100 \text{ liters of HCl}$$
$$\text{molarity} = 2.50\ M$$

 2. What is the molarity of 25.0 m*l* of HCl solution if it is neutralized by 30.0 mL of 5.00 *M* NaOH? _____

An acid and a base neutralize each other to make a salt and water. An example is the reaction of sulfuric acid—H_2SO_4—and barium hydroxide $Ba(OH)_2$. Fig. 35-2 shows $Ba(OH)_2$ neutralizing H_2SO_4. The salt barium sulfate—$BaSO_4$—is forming as a solid at the bottom of the flask.

Fig. 35-2

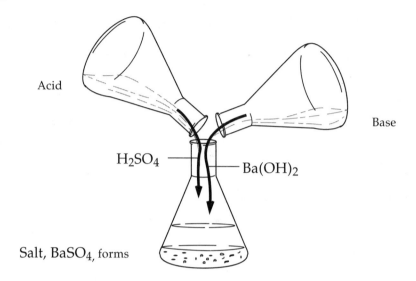

Acid

Base

H_2SO_4 $Ba(OH)_2$

Salt, $BaSO_4$, forms

$$H_2SO_4 + Ba(OH)_2 \longrightarrow BaSO_4 + 2H_2O$$

Fig. 35-3

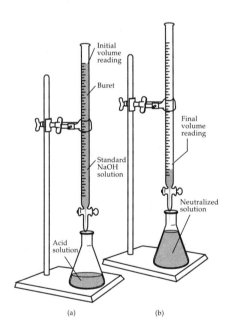

(a)   (b)

Fig. 35-3 shows the procedure for titrating HCl with a standard NaOH solution. A known volume of HCl and phenolphthalein is placed in the flask. A standard NaOH solution is poured into the burette. The volumes of the NaOH solution at the beginning and at the end of the titration are recorded. The volume of NaOH used is found by subtraction.

Use the Key Words from the beginning of this lesson to fill in the blanks.

3. An acid and a base react to form a salt and water in a reaction called

    _____.

4. A compound made of a positive ion other than H^+ and a negative ion
    other than OH^- is a(n) _____.

5. The process used to measure the concentration of an unknown solution
    by reacting it with a standard solution is _____.

6. A solution of known concentration is a(n) _____.

7. The point of titration when chemically equivalent amounts of acid and
    base are present is the _____.

Fill in the blanks with the correct word or answer to the problem.

8. When hydrobromic acid—HBr—neutralizes NaOH, the formula of the

    salt is _____.

9. If the salt formed during neutralization is calcium sulfate—$CaSO_4$, the
    formula of the acid used is _____. The formula of the base

    used is _____.

10. What is the molarity of 40 ml of NaOH if it is completely neutralized
    by 10 ml of 6.0 M HCl? _____

11. How many ml of 2.0 M KOH will be needed to neutralize 30 ml of 0.50
    M HNO_3? _____

12. How many liters of 2.5 M H_2SO_4 are needed to neutralize 2.5 liters of
    5.0 M $Ca(OH)_2$? _____

13. A 30-ml sample of HCl is neutralized by 10 ml of 1.5 M KOH. What is
    the molarity of the HCl? _____

14. If 30 ml of water is added to the HCl in question 13, how much more
    KOH will be needed for complete neutralization?_____

Summary

- A solution is a mixture of substances whose particles are molecular or ionic in size and are evenly spread throughout the mixture.

- Three types of solutions are gas solutions, liquid solutions, and solid solutions. In each type of solution, the solvent is present in a greater amount than the solute or solutes.

- Substances whose water solutions conduct electricity are electrolytes.

- Solubility is the amount of solute dissolved in a specific amount of solvent at a given temperature.

- Molarity is the concentration of a solution in moles per liter of solution. The concentration of a solution in moles per kilogram of solvent is molality.

- The boiling point of a water solution increases by 0.52°C. per molal solution of nonelectrolyte. The freezing point of a solution decreases by 1.86°C per molal solution of a nonelectrolyte.

- The movement of a solvent through a membrane is osmosis. The solvent flows from areas of low concentration to areas of high concentration. The pressure needed to stop osmosis is osmotic pressure.

- Arrhenius defined acids as substances that produce hydrogen ions and bases as substances that produce hydroxide ions in water solution. Bronsted-Lowry acids are defined as proton donors. Bronsted-Lowry bases are proton acceptors. An acid and a base neutralize each other to form a salt and water.

- Related acid-base pairs differ by one proton. Some substances can act as either an acid or a base.

- The ionization constant of an acid—K_a—defines its relative strength. The pH scale shows the concentration of hydronium ions in acid, base, or neutral solutions.

- Titration is used to determine the concentration of an unknown solution by comparing it to a standard solution of known concentration.

For Your Portfolio

1. Use ten or more key words from the unit to write a poem that explains either acid-base reactions or solution chemistry.

2. In business, a ledger is used to show debits and credits. Set up a chemistry ledger to show how proton donors balance proton acceptors.

3. Make a three-dimensional model to show how a solute dissolves in a solvent.

4. Imagine you are a choreographer in the "Acid-Base Theater." Design a dance to show how protons move between related acid-base pairs.

5. Make a videotape of a laboratory exercise that shows boiling point elevations and freezing point depressions.

6. Trace the history of acids either before or after the Industrial Revolution.

In taking this test, you may refer to the tables in Fig. 31-1: *Solubilities in Water* on page 149, Fig. 31-2: *Solubility Curves* on page 150, and Fig. 34-2: *Strengths of Acids* on page 164 when needed.

Fill in the blanks with the correct answers.

1. The temperature at which 60 g of $KClO_3$ will dissolve in 100 g of water is

    _____.

2. A one-molal sugar solution has a freezing point that is less less than 0°C and a

    boiling point that is _____ than 100°C.

3. The number of grams of KCl per 100g of water that would produce a saturated

    solution at 50°C is _____.

4. Molarity expresses concentration in units of _____,

    while molality expresses concentration in units of _____.

5. According to the Bronsted-Lowry theory, an acid is a(n) _____.

6. According to the Arrhenius theory, a solution that forms OH^- ions as the only

    negative ion is a(n) _____.

7. Pure water contains equal concentrations of H_3O^+ and _____.

8. The number of m*l* of 8.00 *M* NaOH required to neutralize 200 m*l* of 4.00 *M* HCl is

    _____.

9. The pH value of a solution in which the $[H_3O^+]$ is 1×10^{-4} is _____.

Answer one of the following questions. Use complete sentences in your answer.

10. **a.** What is one way in which a solution of sugar in water and a solution of carbon dioxide in water are similar? What is one way other than having different substances as the solute in which these solutions differ?

    b. Why does the freezing point of a one-molal solution of sodium chloride depress the freezing point of water by twice as much as a one-molal solution of sugar?

Energy and Chemical Reactions

Take a deep breath. Now think about the air that you took into your body. How free of pollutants was that air? If you are anywhere near cars giving off exhaust, the air may not be very clean. It may contain toxic gases such as carbon monoxide, nitrogen oxides, and unburned fuel. But the air you breathe has less pollutants than it would have if there were no catalytic converters.

What's a catalytic converter? It's a device containing a substance known as a catalyst that helps some chemicals change to others. Most converters have two different catalysts that contain metals such as platinum. One catalyst oxidizes carbon monoxide to carbon dioxide. This catalyst also helps convert unburned fuel to carbon dioxide and water. A second catalyst changes nitrogen oxides to nitrogen gas. In this unit, you'll find out about the role of catalysts as well as other factors that affect chemical reactions.

Lesson 36 Reaction Rates

People attempt to change the speed at which reactions take place in many situations. The spoiling of food and the rusting of cars involve reactions that people try to slow down. On the other hand, people try to speed up reactions, such as the setting of dental fillings.

Kinetics. The area of chemistry dealing with the speed at which reactions take place is **kinetics** (kih-NEHT-ihks). The speed of a reaction depends upon two or more molecules colliding in such a way that old bonds break and new bonds form. The **reaction rate** (ree-AK-shuhn rayt) is the change of concentration of reactants in a unit of time. The nature of the reactants, temperature, surface area, concentration, and use of catalysts are all things that affect reaction rates. Let's consider each of these things.

Nature of the Reactants. The materials used in a chemical reaction often determine its rate. For example, hydrogen and oxygen combine quickly and explosively to form water. However, when oxygen reacts with iron to form rust, the reaction takes place slowly.

Temperature. As temperature increases, the average kinetic energy, or speed, of reacting particles increases. This results in more collisions between particles. For example, milk sours faster at room temperature than it does in a refrigerator. In general, increasing the temperature 10°C causes the rate of reaction to double.

Surface Area. Reactions take place when surfaces are in contact. Crushing a solid reactant into small pieces is a way to increase its surface area. More surface area allows more contact between the reactants, and a faster reaction takes place. There is more surface area in a steel wool pad than in a bar of iron, and so the steel wool rusts more quickly.

Concentration. Increasing the concentration of one or more reactants increases the number of possible collisions between the reacting particles. This can speed a chemical reaction. Thus, the greater the concentration of reactants, the greater the reaction rate. In a glowing wooden splint, carbon in the wood is reacting slowly with oxygen of the air. But when the splint is put into a tube containing a greater concentration of oxygen, the burning reaction proceeds fast enough to cause the splint to burst into flame.

Catalysts. A **catalyst** (KAT-uh-lihst) speeds a chemical reaction without itself being permanently changed. Sometimes the catalyst is in a phase that differs from the reacting substances. In a catalytic converter, solid platinum is used as a catalyst to speed gaseous reactions. When the reaction is completed, the platinum is recovered and used again.

In some reactions, the catalyst is in the same phase as the reacting substances. This type of catalyst is first used as a reactant and then released as a product. In the atmosphere, nitric oxide—NO—is a catalyst that speeds the decomposition of ozone—O_3.

The following reactions show how nitric oxide—NO—is first a reactant and finally a product. The catalyst NO can then be reused.

$$NO(g) + O_3(g) \longrightarrow NO_2(g) + O_2(g)$$

$$NO_2(g) + O(g) \longrightarrow NO(g) + O_2(g)$$

1. A substance that speeds a reaction without itself being changed is called a(n) _____.

Reactions in Steps. Often a reaction takes place in a series of steps. The **rate determining step** is the slowest step in a reaction. No matter how fast a separate step may be, the overall reaction can proceed no faster than its slowest step.

For example, there are three steps in a reaction in which A, B, and C react to form D. In Step 1, a molecule of A reacts with B forming X. This reaction is fast. In Step 2, another molecule of A reacts with X to form Y. This reaction is slow. In Step 3, C reacts with Y to form D. This reaction is fast.

Step 1: A + B $\longrightarrow$ X (fast speed)

Step 2: A + X $\longrightarrow$ Y (slow speed)

Step 3: C + Y $\longrightarrow$ D (fast speed)

The overall reaction is: 2A + B + C $\longrightarrow$ D (slow speed)

Step 2, which is the slowest, determined the speed of the overall reaction.

 **2. In a reaction that occurs in steps, the slowest step is called the
_____ step.**

The greater the concentration of reactants, the faster the rate of the reaction. Hydrogen—H_2—and iodine—I_2—react to form hydrogen iodide—HI:

$$H_2 + I_2 \longrightarrow 2HI$$

Look at Fig. 36-1. The drawing at the left shows that if there was just one molecule of each reactant, there would be one possible collision between them. The drawing at the right shows how doubling the number of molecules of the reactants quadruples the number of possible collisions. As a result, the two substances react four times faster when their concentrations are doubled.

Fig. 36-1

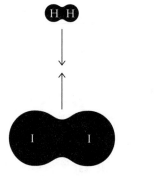 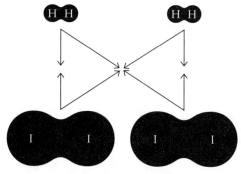

One possible collision between molecules

Four possible collisions among molecules
Reaction rate quadruples

Check Your Understanding

Explain how the following phrases are related.

3. *rate determining step, reaction rate* _____

_____

_____

Complete the statements by adding the correct word or definition.

4. The area of chemistry concerned with the speed of reactions is

_____.

5. The rate of reaction is the _____
   _____.

6. Factors that affect the rate of reaction are

   (a) _____, or substances used;

   (b) _____, which can increase molecular speed;

   (c) _____, which can allow more contact of reactants;

   (d) _____, which brings particles closer together and allows a greater number of collisions between particles; and

   (e) _____, which are not permanently changed.

If the statement is correct, write the word *True*. If the statement is incorrect, write the word *False*.

7. _____ The rate determining step is the fastest step of a chemical reaction.

8. _____ A catalyst speeds a chemical reaction.

9. _____ Increasing the temperature 10°C causes the rate of reaction to double.

10. _____ Reactions take place at surfaces.

11. _____ The change of concentration of reactants in a unit of time is kinetics.

12. _____ A chemical reaction can proceed no faster than its slowest step.

13. _____ Doubling the concentration of reactants can cause the rate of reaction to be quadrupled.

14. _____ All chemical reactions take place at the same speed.

Reaction Equilibrium

Key Words

Haber process:	method of producing ammonia from hydrogen and nitrogen
reversible reaction:	reaction in which the products can react to produce the original reactants
chemical equilibrium:	state in which the forward and reverse reactions of a reversible reaction proceed at the same rate
equilibrium constant, or K_e:	ratio of the concentrations of the products to the reactants in a reversible reaction at equilibrium

KEY IDEAS

Some reactions reach a point at which the products react to form the reactants. The concentrations of reactants and products do not change. This state results when the rate at which reactants form products equals the rate at which the products form the reactants.

Plants require large amounts of nitrogen, but they can't use nitrogen as it occurs in the air. One of the most important chemical processes ever controlled is the reversible reaction used in the synthesis of ammonia—NH_3—from the nitrogen in air. The ammonia is used to make fertilizers that contain nitrogen compounds. These fertilizers are used with plants that are a source of food for people.

Reversible Reactions. Businesses that make nitrogen compounds, such as fertilizers, use the **Haber process** (HAY-ber PRAH-sehs). This process produces commercial amounts of ammonia—NH_3—from hydrogen—H_2—and nitrogen—N_2. The equation for the Haber process is as follows:

Equation 1 $N_2 \ (g) \ + \ 3 \ H_2 \ (g) \rightleftharpoons 2 \ NH_3 \ (g)$

The two arrows show that both reactions take place at the same time. In the forward reaction, indicated by the arrow from left to right ($\rightarrow$), N_2 and H_2 form NH_3. In the reverse reaction, indicated by the arrow from right to left ($\leftarrow$), NH_3 forms N_2 and H_2.

The reaction in Equation 1 is reversible. In a **reversible reaction**, (rih-VER-suh-buhl) the products can react to produce the original reactants. In the Haber process, the original reactants are hydrogen and nitrogen, and the original product is ammonia. Since this product breaks down to form the original reactants, the Haber process is a reversible reaction. That is, hydrogen and nitrogen form ammonia, and at the same time, ammonia forms hydrogen and nitrogen.

Equilibrium. When the forward and reverse reactions of a reversible reaction take place at the same rate, **chemical equilibrium** (ee-kwih-LIHB-ree-uhm) is reached. At equilibrium, the concentration of each substance in the reaction remains the same. Also, the rate of the forward reaction equals the rate of the reverse reaction.

A general equation for any reversible reaction is the following:

Equation 2 $$A + B \rightleftharpoons C + D$$

In this equation, Rate 1 (of the forward reaction) = Rate 2 (of the reverse reaction). Also, each rate is equal to the concentration in moles per liter times a constant. Recall that brackets around a formula represent concentration of the substance in moles per liter. In Equation 3, K stands for a rate constant.

Equation 3 $$K[A][B] = K[C][D]$$

Equilibrium Constant. At equilibrium, the ratio of the concentrations of the products to the reactants has a number value called the **equilibrium constant**, or K_e. For the general reaction in Equation 2 and Equation 3,

Equation 4 $$K_e = \frac{[C][D]}{[A][B]}$$

When there are two or more substances on a side of the equation, their concentrations are multiplied together. Also, the coefficients of the substances in the chemical equation appear as exponents in the equation for the equilibrium constant. See equation 5 and Equation 6.

The chemical equation for the Haber process is in Equation 1. The equilibrium constant for the reaction is written as follows:

Equation 5 $$K_e = \frac{[NH_3]^2}{[N_2][H_2]^3}$$

For another example, K_e for the reaction $H_2 + I_2 \rightleftharpoons 2HI$, is written as:

Equation 6 $$K_e = \frac{[HI]^2}{[H_2][I_2]}$$

1. Write the K_e for $2CO_2 \rightleftharpoons 2\,CO + O_2$

_____

The concentrations of the reactants are written in moles per liter. K_e, however, is expressed without units. Once determined, the value of the equilibrium constant does not change unless there is a change of temperature.

The value of K_e shows how completely the original reactants have been changed to products. A large K_e, or a value greater than 1, means that a large amount of products form from the reactants in the forward reaction. A small K_e, or a value less than 1, means that only a small amount of the reactants change to products.

Another way of looking at the Haber process is shown in the graph in Fig. 37-1. It shows nitrogen and hydrogen combining to form ammonia. Note that at the beginning of the reaction there is an abundance of nitrogen and hydrogen and no ammonia. The reaction reaches equilibrium when the concentrations of all three components remain constant.

Fig. 37-1

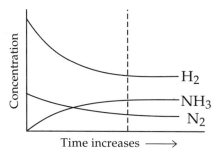

Check Your Understanding

2. Write sentences that show how the terms _chemical equilibrium_, _reversible reaction_ and K_e are related. _____

_____

_____

_____

Complete the statements by writing the correct word in each blank.

3. The _____ process is a method to prepare ammonia.

4. _____ occurs in a reversible reaction when forward and reverse reaction rates are equal.

5. In a(n) _____ the products can react to produce the original reactants.

6. The _____ is the ratio that compares the concentration of products to the concentration of the reactants.

7. At equilibrium, the ratio of the concentrations of products to reactants has a number value with the symbol _____.

8. A large K_e means that a large amount of _____ are formed.

9. A(n) _____ K_e indicates that only a small amount of reactants are changed to products.

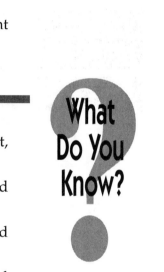

If the statement is correct, write the word *True*. If the statement is incorrect, write the word *False*.

10. _____ The products of the Haber process are hydrogen and oxygen.

11. _____ In a reversible reaction, the products of the forward reaction can react to produce the original reactants.

12. _____ When chemical equilibrium is reached, the forward reaction takes place faster than the reversible reaction.

13. _____ Equilibrium constant is abbreviated as K_e.

14. _____ The value of an equilibrium constant does not change unless there is a change of concentration.

15. _____ In an equilibrium constant expression, coefficients of the formulas in the equation become exponents of the concentrations.

16. _____ A large equilibrium constant indicates that small amounts of products are formed.

38 Influencing Reaction Equilibrium

Le Châtelier's principle:	principle stating that if a reversible reaction at equilibrium is stressed, the equilibrium will shift to relieve the stress
equilibrium stresses:	changes that alter the reaction rate, adjusting the direction of reaction movement
endothermic reaction:	reaction that absorbs heat energy
exothermic reaction:	reaction that releases heat energy

KEY IDEAS

Some reactions proceed in the forward direction and do not reverse. Other reactions may reach equilibrium. Chemical reactions at equilibrium can be forced to move in a forward or a reverse direction by a change in concentration, temperature, or pressure.

While it is difficult to make nitrogen compounds from atmospheric nitrogen, making other nitrogen compounds from ammonia is relatively easy. Le Châtelier's principle is applied in the Haber process, which is used to manufacture ammonia from atmospheric nitrogen and hydrogen. The following table shows some useful nitrogen compounds made from ammonia.

Nitrogen Compounds

Compound	Formula	Uses
Nitric Acid	HNO_3	Explosives, fertilizer
Hydrazine	N_2H_4	Rocket fuel
Nitrates	$M\text{-}NO_3$ ex: sodium nitrate ($NaNO_3$)	Food preservative, fertilizer, explosives
Nitrites	$M\text{-}NO_2$	Food preservative, bleach
Urea	$(NH_2)_2CO$	Fertilizer

Reactions Go to Completion.   A reversible reaction has two directions, forward and reverse. When the products of the forward reaction remain in the reaction mixture, they often react to reform the original reactants. However, if a product of the forward reaction is removed from the reaction mixture, the forward reaction goes to completion. Removing a product prevents it from reacting to reform reactants. There is no reverse reaction. There are three common ways that products are removed so that a reaction goes to completion.

Formation of a Precipitate.   A product that is an insoluble substance, or precipitate, settles to the bottom of the reaction container. This substance does not dissolve, so it does not return ions to the solution. The reaction stops because the ions are not available.

For example, silver ions from $AgNO_3$ and bromide ions from KBr join to form AgBr, which is insoluble. Thus, the reaction goes to completion.

Equation 1 $AgNO_3$ *(aq)* + KBr *(aq)* —> KNO_3 *(aq)* + AgBr *(s)*

Formation of a Gas.   If one product of a reaction is a gas, it can bubble out of the reaction container. The reaction stops because the ions that formed the gas are not available For example, hydrogen ions from HCl and sulfide ions from FeS join to form H_2S gas, which bubbles out of the solution. Thus, the reaction goes to completion.

Equation 2   FeS *(s)* + 2HCl *(aq)* —> $FeCl_2$ *(aq)* + H_2S *(g)*

Formation of Water.   Water that forms in a neutralization reaction is only slightly ionized. Thus, the hydrogen ion and the hydroxide ions are removed, and the reaction goes to completion. For example, hydrogen ions from HCl and hydroxide ions from NaOH join to form water, that is slightly ionized. Thus, the reaction goes to completion.

Equation 3   HCl *(aq)* + NaOH *(aq)* —> NaCl *(aq)* + H_2O*(l)*

   1.   Three kinds of products that cause a reaction to go to completion are

_____

_____.

Shifting Equilibrium.   **Le Châtelier's principle** (luh-shah-tehl-YAYZ) states that if a reversible reaction at equilibrium is stressed, the equilibrium will shift to relieve the stress. Changes of concentration, pressure, and temperature are three stresses that can be applied to an equilibrium system.

If the equilibrium shifts to favor the forward reaction, the rate of the forward reaction increases, and more of the products of the forward reaction are formed. If the equilibrium shifts to favor the reverse reaction, then the rate of that reaction increases. In the Haber process, applying stresses that increase the rate of the forward reaction increases the amount of ammonia produced. The **equilibrium stresses** (ee-kwih-LIHB-ree-uhm STREH-suhs) that can alter the reaction rate and adjust the direction of the reaction can be of three types.

Change of Concentration. Increasing the concentration of a reactant, or decreasing the concentration of a product, is one stress that can be applied to a reaction. This stress forces the reaction toward the right. Look at the equation for the Haber process:

Equation 4 $\qquad$ $N_2\ (g)\ +\ 3H_2\ (g) \rightleftharpoons 2NH_3\ (g)$

Increasing the concentration of N_2 or of H_2 shifts the equilibrium to the right, and more NH_3 is produced. Decreasing the concentration of NH_3 also shifts the equilibrium to the right.

 2. Which concentration has to be increased to shift the equilibrium to the left? _____

Change of Pressure. Increasing pressure forces the reaction toward the smaller volume. When all the substances in a reaction are gases, their volumes are in the same ratio as their coefficients in the equation. On the left side of Equation 4, the coefficients are 1 for N_2 and 3 for H_2, and their sum is 4. On the right side, the coefficient of $2NH_3$ is 2. Thus, there are 4 volumes of reactants and 2 volumes of product, and increasing the pressure forces the reaction toward the right.

Change of Temperature. An **endothermic reaction** (ehn-doh-THER-mihk) absorbs heat energy. An **exothermic reaction** (ehk-soh-THER-mihk) releases heat energy. When the energy change is added to Equation 4 for the Haber process, it appears on the right side, like this:

Equation 5 $\qquad$ $N_2(g)\ +\ 3H_2(g) \rightleftharpoons 2NH_3(g)\ +$ heat

This shows that the reaction is exothermic from left to right and endothermic from right to left. So raising the temperature or adding heat forces the reaction to the left.

Fig. 38-1

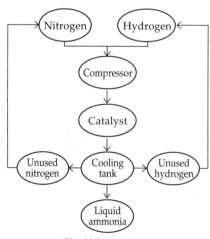

The Haber process

The flowchart in Fig. 38-1 on p. 186 shows the Haber process, which is a reversible reaction at equilibrium. The hydrogen and nitrogen gases are compressed at pressures up to 1000 atmospheres. This increased pressure causes the equilibrium in Equation 5 to shift toward the right. A catalyst of iron and metallic oxides is also used to speed the reaction rate. Unused hydrogen and nitrogen are recycled.

Check Your Understanding

3. How are the terms *exothermic* and *endothermic* related? _____

_____

_____

_____

Fill in the correct words in the statements.

4. A reaction goes to _____ when a product is removed.

5. Products are removed from the reaction mixture by the formation of a

   (a) _____, **(b)** _____, or **(c)** _____.

6. The _____ principle states that if an equilibrium is stressed, the equilibrium will shift to relieve the stress.

7. Three stresses that can be applied to an equilibrium are changes of
   (a) _____, **(b)** _____, or **(c)** _____.

If the statement is correct write the word *True*. If the statement is incorrect write the word *False*.

8. _____ Removing a product from a reaction blocks the reversal of the reaction.

9. _____ Formation of a gas allows a reaction to go to completion.

10. _____ Increasing the pressure on a reaction at equilibrium always shifts the equilibrium toward the reactants.

11. _____ An endothermic reaction releases heat.

12. _____ Removing heat from a reaction at equilibrium favors an exothermic reaction.

What Do You Know?

Lesson 39 Heat of Reaction

Key Words

activated complex:	temporary group of atoms that form as reactant particles rearrange to form products
activation energy:	energy needed to move reactants into the activated complex
enthalpy, or H:	total energy content of a substance
entropy, or S:	measure of randomness or disorder
spontaneous reaction:	reaction that takes place without the addition of outside energy
free energy, or G:	indication of how spontaneous a reaction is, as determined by the effects of heat, temperature, and entropy

KEY IDEAS

Energy must be applied to make reacting substances come together, rearrange, and form products. Energy is absorbed or given off in these chemical reactions.

In the body, enzymes are catalysts. Some of these catalysts reduce the amount of energy needed to start reactions. The catalysts speed reactions that release energy needed to sustain life. Outside the body, sugar does not readily react with oxygen. However, in the presence of enzymes in the body, sugar reacts easily, releasing carbon dioxide, water, and the energy needed for life functions.

Activation Energy. When hydrogen and oxygen are mixed at room temperature, no reaction takes place. However, if a flame is inserted into the mixture of these gases, they react rapidly and form water. The flame provides enough energy for the molecules of hydrogen and oxygen to collide. During the collisions, existing bonds break and new bonds form.

The flame in this example supplies the energy needed to move reactants into a temporary group of atoms called an **activated complex** (ak-tuh-VAYT-uhd KAHM-plehks). The activated complex forms as reactant particles rearrange to form products. The amount of energy that must be applied to the reactants, allowing them to shape an activated complex, is called the **activation energy** (ak-tuh-VAY-shuhn) .

Consider the example of hydrogen and oxygen forming water. For each mole of water that is formed, 57.8 kilocalories of heat are given off. Recall that coefficients in an equation indicate the relative numbers of moles of the substances. Dividing each coefficient by 2, you can write the equation as follows to show the amount of heat given off for 1 mole of the product H_2O:

$$H_2 \, (g) \; + \; \tfrac{1}{2} O_2 \, (g) \longrightarrow \; H_2O \, (g) \; + \; 57.8 \text{ kcal}$$

Fig. 39-1 shows the energy diagram for the reaction between hydrogen and oxygen.

Fig. 39-1

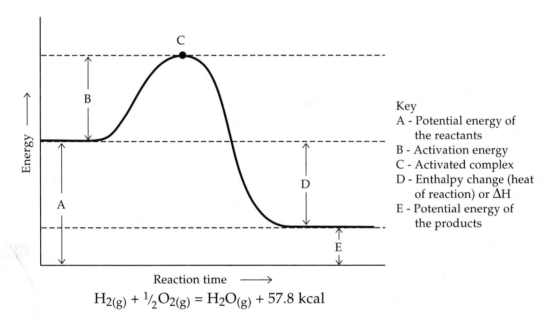

Key
A - Potential energy of the reactants
B - Activation energy
C - Activated complex
D - Enthalpy change (heat of reaction) or ΔH
E - Potential energy of the products

$$H_{2(g)} + \tfrac{1}{2}O_{2(g)} = H_2O_{(g)} + 57.8 \text{ kcal}$$

Follow the diagram from left to right. Use the key to identify A, B, C, D, and E on the diagram. At the beginning of the reaction hydrogen and oxygen have a certain amount of potential energy (A). More energy (B) is applied to the reactants until they reach an activated complex (C). The atoms realign, release energy, and form water. The energy released is the heat of reaction (D). The product water has less potential energy (E) than the reactants hydrogen and oxygen.

Enthalpy. The total energy content of a substance is **enthalpy** (EHN-thal-pee), or **H**. The heat given off or absorbed during a chemical reaction is a change in enthalpy, or ΔH. The synthesis of water is an exothermic reaction, in which heat is given off. Fig. 39-1 shows that ΔH, also called the heat of reaction, is the difference between the potential energies of the product H_2O and the reactants H_2 and O_2. Using the key in Fig. 39-1, this difference can be written as E–A.

Fig. 39-2

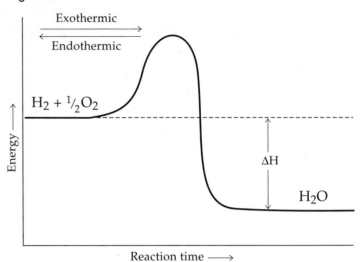

Fig. 39-3

Compound	(ΔH) kcal/mol
Aluminum oxide Al_2O_3 (s)	−400.5
Ammonia NH_3 (g)	−11.0
Carbon dioxide CO_2 (g)	−94.1
Ethane C_2H_6(g)	−20.2
Ethene (ethylene) C_2H_4 (g)	12.5
Magnesium oxide MgO (s)	−143.8
Nitrogen (II) oxide NO (g)	21.6
Sodium chloride NaCl (s)	−98.3
Water H_2O (g)	−57.8

Fig. 39-2 lists values of ΔH for the formation of some compounds. A minus sign indicates an exothermic reaction, in which heat is given off. This means that the product has less enthalpy than the reactants. Fig. 39-3 shows energy change in both the decomposition and the synthesis of water. Any reversible reaction is exothermic in one direction and endothermic in the other direction.

Effect of a Catalyst. The addition of a catalyst to reactants lowers the activation energy. In Fig. 39-4, a dashed line shows the effect of the catalyst upon a reaction. You can see in the diagram that the activation energy is the only thing that changes.

Fig. 39-4

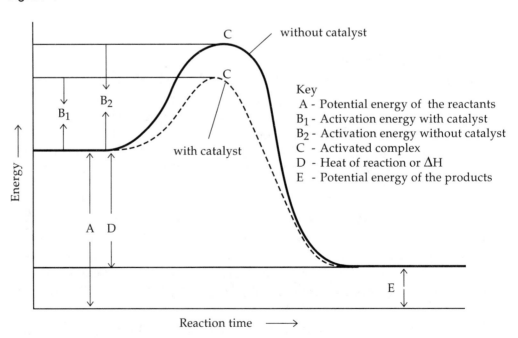

Key
A - Potential energy of the reactants
B_1 - Activation energy with catalyst
B_2 - Activation energy without catalyst
C - Activated complex
D - Heat of reaction or ΔH
E - Potential energy of the products

Entropy and Spontaneous Reactions. If an egg falls off a table and hits the floor, it breaks and its disorder, or entropy, increases. It has gone to a state of greater disorder. **Entropy** (EHN-truh-pee) or **S** is a measure of randomness or disorder.

Chemical reactions behave in a similar way to the egg. A reaction tends to move from a point of high energy to a point of low energy. Also, the entropy of a system tends to increase. Reactions that move from high energy to low energy with an increase in entropy are spontaneous. A **spontaneous reaction** (spahn-TAY-nee-uhs) takes place without the addition of outside energy.

Free energy. A reaction is determined to be spontaneous by calculating its free energy. **Free energy,** or **G,** is an indication of how spontaneous a reaction is, as determined by the effects of heat, temperature, and entropy. Free energy is expressed in a relationship called the Gibbs equation:

$$\Delta G = \Delta H - T\Delta S$$

In this equation, ΔG is free energy change, ΔH is change of enthalpy (heat of reaction), T is the Kelvin temperature, and ΔS is the entropy change.

Fig. 39-5 shows free energies for some substances. A reaction is spontaneous if ΔG is negative. If ΔG is positive, the reaction is not spontaneous.

Fig. 39-5

Compound	Free Energy kcal/mol (ΔG)
Aluminum oxide Al_2O_3 (s)	–378.2
Ammonia NH_3 (g)	–3.9
Carbon dioxide CO_2 (g)	–94.3
Carbon monoxide CO (g)	–32.8
Copper (II) sulfate $CuSO_4$ (g)	–158.2
Ethane C_2H_6 (g)	–7.9
Ethene (ethylene) C_2H_4 (g)	16.3
Ethyne (acetylene) C_2H_4 (g)	50.0
Magnesium oxide MgO (s)	–136.1
Nitrogen (II) oxide NO (g)	20.7
Nitrogen (IV) oxide NO_2 (g)	12.3
Potassium chloride KCl (g)	–97.8
Sodium chloride $NaCl$ (s)	–91.8
Sulfur dioxide SO_2 (g)	–71.7
Water H_2O (g)	–54.6
Water H_2O (l)	–56.7

1. **Is the reaction to produce water spontaneous?** _____

2. **Is the formation of ethene (ethylene) spontaneous?** _____

The following table summarizes the components of the Gibbs equation to predict whether or not a reaction will be spontaneous.

ΔH	ΔS	ΔG	Reaction is
+	−	always positive	nonspontaneous at all temperatures
−	+	always negative	spontaneous at all temperatures
+	+	positive at low temperatures; negative at high temperatures	nonspontaneous at low temperatures; spontaneous as temperature increases
−	−	negative at low temperatures; positive at high temperatures	spontaneous at low temperatures; nonspontaneous at high temperatures

TAKE ANOTHER LOOK

Every reaction is exothermic in one direction and endothermic in the other direction. Fig. 39-6 shows a catalyzed reversible reaction.

Fig. 39-6

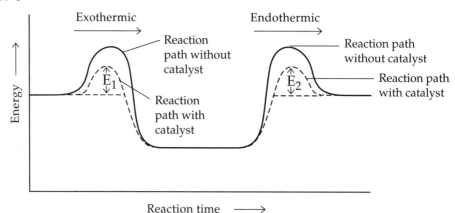

E_1 Activation energy of Exothermic reaction with catalyst

E_2 Activation energy of Endothermic reaction with catalyst

Look at these three things in the diagram:

- The catalyst reduces the amount of activation energy needed.

- The enthalpy decreases in the exothermic reaction, and ΔH has a negative value.

- The enthalpy increases in the endothermic reaction, and ΔH has a positive value.

Check Your Understanding

Complete the following by adding the correct term.

3.   Activation energy raises the reaction to the _____.

4. The heat given off or absorbed during a chemical reaction is

   _____.

5. A measure of disorder is _____.

6. A(n) _____ reaction tends to take place without outside energy and has a negative free energy.

7. The energy needed to move the reactants into the activated complex is

   the _____.

Given symbols, write the terms and definitions in the blanks.

	Term	Symbol	Definition
8.	_____	G or ΔG	_____
9.	_____	H or ΔH	_____
10.	_____	S or ΔS	_____

If the statement is true, write the word *True*. If the statement is false, write the word *False*.

11. _____ Heat is always released during a chemical reaction.

12. _____ A negative ΔH indicates an exothermic reaction.

13. _____ Energy released from chemical reactions is activation energy.

14. _____ Entropy is a measure of disorder.

15. _____ A catalyst slows a reaction by decreasing entropy.

16. _____ A reaction is spontaneous if ΔH is negative and ΔS is positive.

17. _____ The formation of ammonia is an exothermic reaction.

18 _____ Nitrogen (II) oxide—NO—forms spontaneously.

19. _____ Adding a catalyst reduces the heat of reaction.

20. _____ Every reaction is exothermic in one direction and endothermic in the other direction.

21. _____ The difference between the potential energy of the reactants and the potential energy of the products is the activated complex.

Summary

- Kinetics deals with reactions and the rates at which they take place. Reaction rate is the change of concentration in a unit amount of time. Five things affect reaction rates. They are the: nature of the reactants, temperature, surface area, concentration, and catalysts.

- A reaction can be only as fast as its slowest step—the rate determining step.

- A reversible reaction is one in which the products react to reform the reactants. Chemical equilibrium is established in a reversible reaction.

- According to Le Châtelier's principle, a stress can shift an equilibrium. Three stresses that affect equilibria are change of concentration, change of pressure, and change of temperature.

- Reactions go to completion, or do not reverse, when one or more of the products is removed from the reaction mixture. This removal can occur when a product is a gas, a precipitate, or water.

- Catalysts speed chemical reactions. Some catalysts are in the same phase as the reactants. Others are in a phase that differs from the reactants.

- Reactants must collide before they can rearrange to form products. The collision structure is called the activated complex. The energy needed to bring atoms into the activated complex is the activation energy.

- Enthalpy is the energy content of substances. Enthalpy change is the heat absorbed or released in a chemical reaction.

- The amount of enthalpy change is the difference between the potential energies of the reactants and the product. This change is also called heat of reaction, or ΔH.

- Entropy is a measure of disorder. Reactions tend to proceed from a position of high energy to a position of low energy, with an increase in entropy.

- Free energy indicates how spontaneous a reaction is. Free energy can be computed by using the Gibbs equation:

$$\Delta G = \Delta H - T\Delta S$$

If ΔG is negative, the reaction is spontaneous. A spontaneous reaction has a tendency to take place without outside energy.

For Your Portfolio

1. Compare chemical reactions to a thriller novel. Write and act out a skit in which the hero, using activation energy and catalysts, can manipulate the other characters of the novel.

2. Make a model that shows reactants coming together, forming an activated complex, and then forming products.

3. Stresses affect chemical equilibria as well as body movements in athletics. Select one specific sport and show the similarity between chemical reactions and kinesthetics.

4. Compare the tune "The Bear Went over the Mountain" to a potential energy diagram. Prepare a musical score to accompany a graph that never ends.

5. Explain this nursery rhyme "Humpty Dumpty Sat on a Wall" to the class or to an individual by showing how this rhyme is similar to chemical reactions.

In each blank, write the term or expression that correctly completes the statement. Use terms from the following list: *activation energy, heat of reaction, decreases, increases, rate of reaction, potential energy, activated complex,* C_2H_6, ΔG

1. Adding a catalyst to a chemical reaction will cause a change in the _____.

2. The temporary group of atoms that form as reactant particles rearrange to form products is called the _____.

3. In a chemical reaction, the difference between the potential energy of the reactants and the potential energy of the products is _____.

4. A change of free energy is _____.

5. If the concentration of one or more of the reactants is increased, the rate of reaction usually _____.

6. As the surface area of a metal reacting in acid is increased, the rate of reaction usually

    _____.

7. As the temperature decreases, the rate of a reaction usually _____.

8. According to Fig. 39-5, _____ is a compound that forms spontaneously.

Answer one of the following questions.

9. Given the reaction at equilibrium, $2H_2$ (g) $+$ O_2 (g) $\rightleftharpoons$ $2\,H_2O$ (g) $+$ heat.

    Use a complete sentence to describe a change that would force the reaction either to the right or to the left.

10. For Numbers 1 through 5 on the following diagram, write the letter of the correct label: **(a)** activated complex **(b)** effect of catalyst **(c)** activation energy **(d)** potential energy of products **(e)** potential energy of reactants.

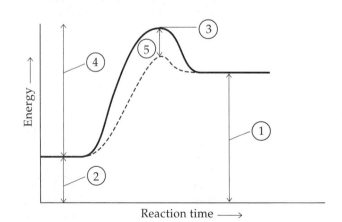

1. _____

2. _____

3. _____

4. _____

5. _____

Electrochemical Reactions

Think about car engines powered by batteries instead of gasoline. Since electric cars do not produce exhaust fumes, how would this affect air pollution? Electric cars have been made for many years, but they are not widely used. A major reason for the limited use of electric cars is that the batteries used to power these cars are very heavy and don't last very long. In recent years, however, new batteries have been developed to overcome these problems.

What inside a battery produces electricity? The answer is chemical reactions that take place inside the battery. In these reactions, electrons move without any outside energy being added.

In other chemical reactions, energy must be added to make electrons move. Such reactions are used in the process of electroplating. In electroplating, metal ions such as gold and silver are plated onto less expensive metals to make jewelry and other products.

In this unit, you will learn about different kinds of electrochemical reactions and how they occur.

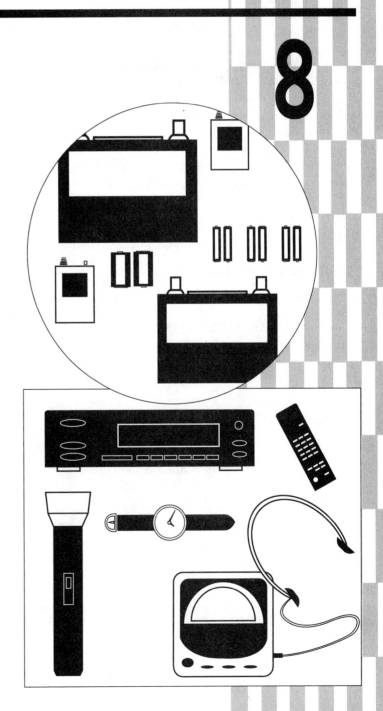

40
Oxidation and Reduction

Key Words

oxidation:	the loss of electrons
reduction:	the gain of electrons
oxidizing agent:	takes or gains electrons from another substance
reducing agent:	gives or loses electrons to another substance

KEY IDEAS

The electrons moving around the nuclei of atoms may also move from one atom to another. Such movement takes place during both oxidation reactions and reduction reactions. Both types of reactions happen at the same time. The loss of electrons is called oxidation. The gain of electrons is called reduction.

Firefighters need to be aware of the processes of oxidation and reduction. When materials burn, they combine with oxygen; thus, oxidation takes place. To prevent or stop fires, firefighters must know how to keep oxygen away from materials that can burn.

Oxidation has two meanings. One meaning is "the combining of a substance with oxygen." For example, calcium combines with oxygen in the oxidation reaction shown below.

$$2\,Ca \ + \ O_2 \longrightarrow 2\,CaO$$

Loss and Gain of Electrons. Another meaning of **oxidation** (ahks-ih-DAY-shuhn) is the "loss of electrons." In the oxidation reaction shown above, the calcium loses electrons. See Fig. 40-1.

Fig. 40-1

$$Ca \longrightarrow Ca^{2+} + 2e^-$$

Calcium	Calcium	2 Electrons
atom	ion	

Calcium can also react with sulfur in an oxidation reaction.

$$Ca + S \longrightarrow CaS$$

In this reaction, the calcium loses electrons just as it does in the calcium-oxygen reaction shown on page 198.

During oxidation, the electrons that are lost go to another substance. For example, electrons lost by calcium can go to sulfur. See Fig. 40-2.

Fig. 40-2   Fig. 40-3

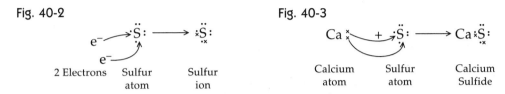

In the calcium-sulfur reaction, shown in Fig. 40-3, the electrons that are lost by the calcium are gained by the sulfur. The gain of electrons by a substance in a reaction is called **reduction** (rih-DUK-shuhn).

Oxidizing and Reducing Agents. In the calcium-sulfur reaction, shown in Fig. 40-3, the sulfur is the **oxidizing agent** (AHKS-ih-dyz-ihng AY-juhnt) because it causes the loss of electrons. The oxidizing agent is reduced because it gains electrons.

The calcium is the **reducing agent** (rih-DOOS-ihng AY-juhnt) because it causes the gain of electrons. The reducing agent is oxidized because it loses electrons.

 1. What is oxidation? _____

_____

 2. What is reduction? _____

_____

Changing Charges. During oxidation reactions, charges on atoms change. Look again at the reaction of calcium and sulfur.

$$Ca^\circ + S^\circ \longrightarrow Ca^{2+}S^{2-}$$

Here is how the charges change: Calcium (Ca°), the reducing agent, becomes more positive (Ca^{2+}); sulfur (S°), the oxidizing agent, becomes more negative (S^{2-}).

To remember what reactions involve losses and gains of electrons, think about this story. There once was a lion named **LEO**, which means **L**oss of **E**lectrons is **O**xidation. This king of the beasts was a little different from others. Instead of saying "roar," he said "**GER**", which means **G**ain of **E**lectrons is **R**eduction. See Fig. 40-4.

Fig. 40-4

Look at Fig. 40-5. It shows the gain and loss of electrons by oxidizing agents and by reducing agents.

Fig. 40-5

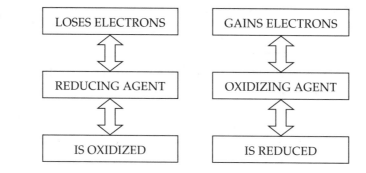

Check Your Understanding

3. Use the terms *reducing agent*, *reduction*, and *substance oxidized* to complete the table comparing oxidation and reduction.

Loss of Electrons	Gain of Electrons
(a) oxidation	(a) _____
(b) _____	(b) oxidizing agent
(c) _____	(c) substance reduced

Fill in the blanks with the proper terms.

Oxidation is the (4)_____ of electrons.   Reduction is the

(5)_____ of electrons. The oxidizing agent (6)_____

electrons. The reducing agent (7)_____ electrons. The charge of

the oxidizing agent becomes more (8)_____. The charge of the

reducing agent becomes more (9)_____.

Two hydrogen atoms and one oxygen atom combine to form water as
shown in Fig. 40-6.

Fig. 40-6

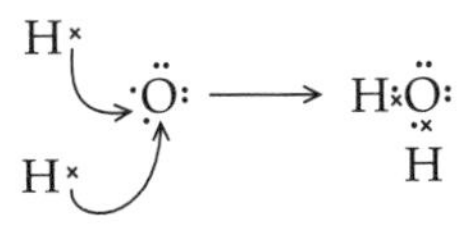

10. Which element is the oxidizing agent? _____

11. Which element is the reducing agent? _____

12. When magnesium burns, it combines quickly with oxygen. When iron
     rusts, the iron reacts slowly with oxygen. Write two equations to show
     that each of these reactions are examples of oxidation.

     _____

     _____

13. Iron ore is reduced to iron in the following reaction. Explain why carbon
     (C) is the reducing agent.

$$2FeO + C \longrightarrow CO_2 + 2Fe$$

14. Hydrogen will react with chlorine as shown in this equation.

$$H_2 + Cl_2 \longrightarrow 2HCl$$

     Write the formula for the oxidizing agent (a)_____, the

     reducing agent (b)_____, the substance oxidized

     (c)_____, and the substance reduced (d)_____.

15. Write a paragraph to explain why oxidation and reduction both take
     place at the same time.

     _____

     _____

     _____

     _____

Lesson 41 Redox Reactions

Key Words

oxidation number: number given to each atom in a chemical formula to show the number of electrons that might be gained, lost, or shared during bond formation.

redox reaction: short term for an oxidation-reduction reaction

KEY IDEAS

In a redox reaction, oxidation numbers change. These numbers are used to show the direction of electron movement in the reactions. When an atom loses electrons, its oxidation number increases. When an atom gains electrons, its oxidation number decreases.

Redox reactions that take place in the body can lead to disease and aging. Antioxidants can stop or slow down harmful redox reactions. For this reason, nurses and other health care workers need to know about antioxidants present in foods and medicines.

Finding Oxidation Numbers. Electrons are gained, lost, or shared when atoms bond together. Oxidation numbers are used to keep track of electrons during bonding. It is easy to find the **oxidation number** of an atom by using the following set of rules:

The oxidation number of a one-atom ion is equal to its charge. For example, the oxidation number of calcium in Ca^{2+} is +2. The oxidation number of sulfur in S^{2-} is –2.

The oxidation number of an element is zero. An uncombined atom such as K or P has an oxidation number of zero. When atoms of the same element bond together, each atom also has an oxidation number of zero. Thus the oxygen atoms in O_2 and the oxygen atoms in ozone O_3 both have oxidation numbers of zero.

In compounds made up of only two elements, the more electronegative element has a negative oxidation number. The less electronegative element has a positive oxidation number. In PCl_3, chlorine is more electronegative than phosphorus. Chlorine therefore has an oxidation number of –1. Phosphorus in PCl_3 is less electronegative than chlorine. Thus, phosphorus has a charge of +3.

In compounds, hydrogen usually has an oxidation number of +1. Oxygen usually has an oxidation number of –2. In HCl, the oxidation number of hydrogen is +1. In CaO, the oxidation number of oxygen is –2.

The sum of the oxidation numbers in an ion made up of many elements is equal to its charge. One example is the nitrate ion NO_3^-, shown in Fig. 41-1. In this ion, each oxygen atom has an oxidation number of –2. Three oxygen atoms have an oxidation number of –6, since 3(–2) = –6. The sum of the oxidation numbers is the charge on the ion, which is –1. That is, the oxidation number of nitrogen added to –6 should equal –1. So the oxidation number of the nitrogen must be +5.

Fig. 41-1

In the sulfate ion SO_4^{2-}, the oxidation numbers add up to –2. Look at Fig. 41-2. Each oxygen atom has an oxidation number of –2. The oxidation number of sulfur is +6 because (+6) + (4)(–2) = –2.

Fig. 41-2

$$\left(S^{6+}O_4^{2-} \right)^{2-}$$

Sulfate ion

The sum of the oxidation numbers in a compound is zero. In water, the oxidation number of the oxygen is –2. The oxidation number of each hydrogen is +1. The oxidation number of both hydrogens is 2(+1) = +2. The sum of –2 for the oxygen and +2 for the hydrogens is zero. In nitric acid HNO_3, the oxidation number of the hydrogen is +1, and the charge on the nitrate ion is -1.

 1. What is the oxidation number of a free element? _____

_____

_____

 2. What is the usual oxidation number of oxygen? _____

_____

_____

 3. What is the sum of the oxidation numbers in a compound?

_____

_____

Oxidation Numbers in Reactions. A **redox reaction** is an oxidation-reduction reaction. Look at the equation shown in Fig. 41-3.

Fig. 41-3

$$\underset{Ba}{\overset{2+}{}}\,\underset{Br_2}{\overset{2(1-)}{}} + \underset{Cl_2}{\overset{0}{}} \longrightarrow \underset{Ba}{\overset{2+}{}}\underset{Cl_2}{\overset{2(1-)}{}} + \underset{Br_2}{\overset{0}{}}$$

In this reaction, the oxidation number of the bromine changes from –1 to 0. The oxidation number of the chlorine changes from 0 to –1. Each bromine atom loses an electron, which is oxidation. Each chlorine atom gains an electron, which is reduction. Thus, the reaction shown is a redox reaction.

Now look at the equation in Fig. 41-4. In this reaction, no change of oxidation numbers occurs. If none of the oxidation numbers change, no redox reaction takes place.

Fig. 41-4

$$\underset{H}{\overset{1+}{}}\underset{Cl}{\overset{1-}{}} + \underset{Na}{\overset{1+}{}}\underset{OH}{\overset{2-1+}{}} \longrightarrow \underset{Na}{\overset{1+}{}}\underset{Cl}{\overset{1-}{}} + \underset{HOH}{\overset{1+2-1+}{}}$$

Look at the redox reaction between sodium (Na) and sulfur (S) shown in Figs. 41-5 and 41-6. The diagrams show the movement of electrons during the same reaction in different ways. Notice that sodium loses electrons, which is oxidation. Sulfur gains electrons, which is reduction. The oxidation number of each sodium atom increases from 0 to +1. The oxidation number of the sulfur atom decreases from 0 to –2.

Fig. 41-5

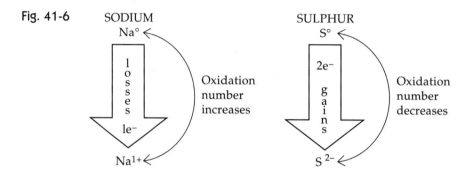

Fig. 41-6

SODIUM

SULPHUR

Check Your Understanding

Fill in the blanks with the correct terms.

The rules for finding oxidation numbers are the following: The oxidation number of an ion equals **(4)**_____. The oxidation number of an element equals **(5)**_____. In two-element compounds, the more electronegative element has a **(6)**_____ charge, and the less electronegative element has a **(7)**_____ charge. Hydrogen usually has an oxidation number of **(8)**_____, and the oxidation number of oxygen is usually **(9)**_____. The sum of the oxidation numbers of an ion equals **(10)**_____. The sum of the oxidation numbers of a compound equals **(11)**_____.

Assign oxidation numbers to each element in the following unbalanced equation.

$$NH_3 + O_2 \longrightarrow NO + H_2O$$

12. The oxidation number of the nitrogen changes from _____ to _____.

13. The oxidation number of the oxygen changes from _____ to _____.

14. What is the oxidation number of each element in the following compounds?

   (a) HCl _____

   (b) H_2SO_4 _____

   (c) $KMnO_4$ _____

   (d) NO _____

15. Which of the following reactions are redox reactions? _____

   (a) $H_2 + Cl_2 \longrightarrow 2HCl$

   (b) $NaCl + AgNO_3 \longrightarrow AgCl + NaNO_3$

   (c) $Zn + CuSO_4 \longrightarrow Cu + ZnSO_4$

   (d) $2HgO \longrightarrow 2Hg + O_2$

Dinitrogen tetroxide (N_2O_4) and hydrazine (N_2H_2) are used as rocket fuels. The reaction between these two compounds produces nitrogen and water, as shown below.

$$N_2O_4 + 2N_2H_4 \longrightarrow 3N_2 + 4H_2O$$

16. What two changes of oxidation number does the nitrogen undergo?

   _____

   _____

17. Does the oxidation number of the oxygen change?_____

   _____

42 Spontaneous Redox Reactions

electrochemical cell:	redox system that produces electricity
half-cell:	metal placed in a solution of a salt of that metal
salt bridge:	tube of salt connecting two half-cells
spontaneous reaction:	reaction that takes place without the use of outside energy
voltage:	measure of electric force

KEY IDEAS

In an electrochemical cell, oxidation and reduction take place in two different half-cells. This reaction is spontaneous, and it produces electricity.

Auto mechanics work with batteries, and batteries are made of electrochemical cells. The reactions inside these cells produce the electricity that starts a car's engine. This electricity also makes the car's lights work and runs the car's other electrical parts.

Redox and Electricity. An **electrochemical cell** (ee-LEHK-troh-KEHM-ih-kuhl) is made of two half-cells, an external conductor, and a salt bridge. These parts are shown in Fig. 42-1.

Fig. 42-1

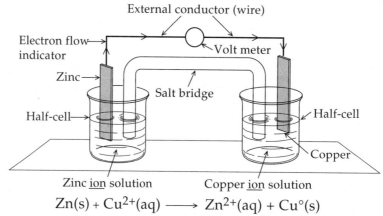

$$Zn(s) + Cu^{2+}(aq) \longrightarrow Zn^{2+}(aq) + Cu°(s)$$

A **half-cell** is made of a metal strip in contact with a solution of its ions. As you can see in the zinc half-cell, the zinc loses electrons. Thus oxidation takes place in the zinc half-cell. Now look at the copper half-cell. Here electrons are gained, so reduction takes place in the copper half-cell.

In Fig. 42-1, the arrows show the path of the electrons. The electrons lost by the zinc metal move into the external conductor, or wire. The electrons then move through the voltmeter and the wire into the copper metal. The copper ions in solution gain the electrons and become copper metal.

The **salt bridge** is a tube that allows ions to move from one solution to another. This bridge completes the circuit. A **spontaneous reaction** (spahn-TAY-nee-uhs) takes place without outside energy. In the cell shown in Fig. 42-1, the reaction goes on until the zinc metal strip wears out. Then the cell no longer works.

Metals and Electrons. Recall that **GER** stands for **G**ain of **E**lectrons is **R**eduction. Thus, a substance with a strong reduction potential is one that has a strong tendency to gain electrons. The table of standard reduction potentials shown in Fig. 42-2 lists substances in order of their tendency to be reduced, or gain electrons.

Voltage (VOHL-tihj) is a measure of electric force. The voltages of reactions taking place in half-cells, or half-reactions, are listed in the table. For example, in a zinc half-reaction, zinc ions (Zn^{2+}) gain electrons to become zinc metal ($Zn°$). The voltage ($E°$) for this half-reaction is -0.76 v. In a copper half-reaction, copper ions (Cu^{2+}) gain electrons to become copper metal ($Cu°$). The voltage for this half-reaction is $+0.34$ v.

All the reactions in the table show reduction, which is a gain of electrons. To find the voltage for oxidation, which is a loss of electrons, turn the equation around. Then reverse the sign of the voltage, as shown below.

reduction: $Zn^{2+} + 2e^- \longrightarrow Zn(s)$ $E° = -0.76$ v

oxidation: $Zn(s) \longrightarrow Zn^{2+} + 2e^-$ $E° = +0.76$ v

Look again at the Cu-Zn cell shown in Fig. 42-1 on page 206. You have already found the oxidation voltage for zinc in this cell. Now use the table to look up the reduction voltage for copper. Add the two voltages together as shown below. The sum is the voltage for the whole cell.

oxidation: $Zn(s) \longrightarrow Zn^{2+} + 2e^-$ $E° = +0.76$ v

reduction: $Cu^{2+} + 2e^- \longrightarrow Cu(s)$ $E° = +0.34$ v

Sum = $+1.10$ v

Fig. 42-2

STANDARD REDUCTION POTENTIALS	
Solutions are 1 M (ag)	
Half-cell Reaction	$E°$ (volts)
$Au^{3+} + 3e^- \blacktriangleright Au(s)$	+1.50
$Hg^{2+} + 2e^- \blacktriangleright Hg(\ell)$	+0.85
$Ag^+ + e^- \blacktriangleright Ag(s)$	+0.80
$Fe^{3+} + e^- \blacktriangleright Fe^{2+}$	+0.77
$Cu^+ + e^- \blacktriangleright Cu(s)$	+0.52
$Cu^{2+} + 2e^- \blacktriangleright Cu(s)$	+0.34
$2H^+ + 2e^- \blacktriangleright Cu(s)$	0.00
$Pb^{2+} + 2e^- \blacktriangleright Pb(s)$	−0.13
$Sn^{2+} + 2e^- \blacktriangleright Sn(s)$	−0.14
$Ni^{2+} + 2e^- \blacktriangleright Ni(s)$	−0.26
$Co^{2+} + 2e^- \blacktriangleright Co(s)$	−0.28
$Fe^{2+} + 2e^- \blacktriangleright Fe(s)$	−0.45
$Cr^{3+} + 3e^- \blacktriangleright Cr(s)$	−0.74
$Zn^{2+} + 2e^- \blacktriangleright Zn(s)$	−0.76
$Al^{3+} + 3e^- \blacktriangleright Al(s)$	−1.66
$Mg^{2+} + 2e^- \blacktriangleright Mg(s)$	−2.37
$Na^+ + e^- \blacktriangleright Na(s)$	−2.71
$Ca^{2+} + 2e^- \blacktriangleright Ca(s)$	−2.87
$Sr^{2+} + 2e^- \blacktriangleright Sr(s)$	−2.89
$Ba^{2+} + 2e^- \blacktriangleright Ba(s)$	−2.91
$Cs^+ + e^- \blacktriangleright Cs(s)$	−2.92
$K^+ + e^- \blacktriangleright K(s)$	−2.93
$Li^+ + e^- \blacktriangleright Li(s)$	−3.04

Fig. 42-3

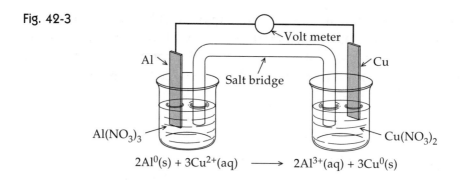

$$2Al^0(s) + 3Cu^{2+}(aq) \longrightarrow 2Al^{3+}(aq) + 3Cu^0(s)$$

Use the table on page 207 to find the voltages for each half-cell of the aluminum-copper cell shown in Fig. 42-3. The reduction voltage for the Al half-cell is –1.66 v. However, the Al will lose electrons because it is listed lower in the table than Cu. The loss of electrons from Al is oxidation.

The voltage for the Cu half-cell is listed as +0.34 v. To find the voltage for this Al-Cu cell, first change the reduction voltage to oxidation for the Al half-cell. Then add the two voltages together as shown below.

oxidation: $Al(s) \longrightarrow Al^{3+} + 3e^-$ $E° = +1.66$ v

reduction: $Cu^{2+} + 2e^- \longrightarrow Cu(s)$ $E° = +0.34$ v

                              Sum $= +2.00$ v

 1. In a copper half-cell connected to a zinc half-cell, which metal loses electrons? _____

 2. In a copper half-cell connected to an aluminum half-cell, which metal gains electrons? _____

 3. What is the voltage of a zinc-silver cell in which the oxidation voltage (for zinc) is +0.76 v and the reduction voltage (for silver) is +0.80 v?

_____

Metal Activity. The table on page 207 can be used to predict replacement reactions between metals and the ions of other metals. The most active metal, lithium (Li), has the lowest tendency to gain electrons, or to be reduced (**GER**). Thus Li is at the bottom of the table. Lithium also has the strongest tendency of the metals listed to lose electrons, or to be oxidized (**LEO**). The least active metal, gold (Au), is at the top of the table, so it has the greatest tendency to be reduced (**GER**). Gold also has the lowest tendency to lose electrons, or to be oxidized (**LEO**).

In a replacement reaction (see Lesson 27), a more active metal replaces one that is less active from a solution of its ions. A metal that is listed lower in the table of reduction potentials is more active, so it can replace any metal that is higher. Magnesium is more active than nickel. Therefore, magnesium will replace nickel from a solution of its ions as shown here.

$$2\,Mg \;+\; Ni(NO_3)_2 \longrightarrow 2\,MgNO_3 \;+\; Ni$$

Study the equation below and the table of standard reduction potentials on page 207. Notice that zinc is lower in the table and will replace the copper from solution.

$$Zn \;+\; CuSO_4 \longrightarrow ZnSO_4 \;+\; Cu$$

4. If an iron nail is placed in a copper sulfate solution, will a reaction take place? Use the table to find out. _____

_____

Look at the equation below.

$$Ag \;+\; CuSO_4 \longrightarrow \text{no reaction}$$

Silver is above copper in the table. Thus silver is less active than copper. A spontaneous reaction will not take place.

Reactions with Acids. Solutions of acids produce hydrogen ions (H^+). All the reduction potentials listed in the table are determined by comparison with the H^+ half-cell, which has a voltage ($E°$) of 0.00 v. Thus the table can be used to predict which metals will react with acids such as sulfuric acid (H_2SO_4) and hydrochloric acid (HCl). Metals listed in the table below hydrogen react spontaneously with these acids. For example, aluminum is listed in the table below hydrogen, so aluminum will react with hydrochloric acid.

Look at the equation below.

$$2Al \;+\; 6HCl \longrightarrow 2AlCl_3 \;+\; 3H_2$$

Find iron in the table. Notice that iron will also react with HCl. Gold is in the table above hydrogen, so gold will not react with H_2SO_4.

5. Will silver (Ag) react with H_2SO_4? _____

_____

Look at Fig. 42-4 which shows a strip of zinc placed into a copper sulfate solution. Copper forms a coating on the zinc. The zinc replaces the copper from solution because zinc is more active than the copper. Thus, a spontaneous redox reaction is taking place.

Look at Fig. 42-5 which shows a strip of magnesium placed into hydrochloric acid. The acid gives off bubbles of hydrogen gas. Magnesium is an active metal listed in the table on page 207 below hydrogen. Thus, magnesium replaces the hydrogen from the acid.

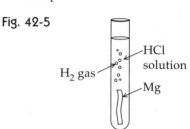

Fig. 42-4

Zinc strip

Copper forming on zinc strip

$CuSO_4$ solution

Fig. 42-5

H_2 gas

HCl solution

Mg

Check Your Understanding

Fill in the blanks with the following terms: *half-cell(s), external conductor, salt bridge; spontaneous, voltage*. The underlined terms will be used more than once.

Electrochemical cells are made of two **(6)**_____, a(n)

(7)_____, and a(n) **(8)**_____. A(n)

(9)_____ reaction takes place without outside energy. A(n)

(10)_____ is a tube of salt connecting two half-cells. A measure

of electric force is **(11)**_____. A(n) **(12)**_____

contains a metal placed in a solution of ions of that metal.

Use Fig. 42-6 to complete items 13 – 17. Lable the half cells in items 13 and 14.

Fig. 42-6

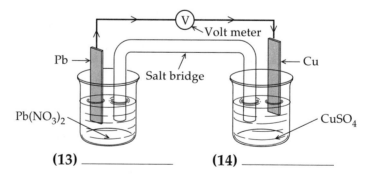

Pb

Pb(NO$_3$)$_2$

Salt bridge

V

Volt meter

Cu

CuSO$_4$

(13) _____   **(14)** _____

15. In which half-cell is there a loss of electrons? _____

16. In which half-cell is there a gain of electrons? _____

17. On the lines below, calculate E° for the cell.

_____

_____

In items 18-21, underline the correct answer.

18. Which metal will replace nickel from solution? Use the table on page 207 to find out. (copper, lead, silver, iron)

19. In an electrochemical cell, ions move through the (salt bridge, external conductor, metal strip, voltmeter).

20. What is the voltage of a cell made of a zinc half-cell and a copper half-cell? (0.55 v, 1.10 v, 1.35 v, 1.85 v)

21. Of the following, which is the most active metal listed in the table on page 207? (gold, hydrogen, zinc, lithium)

Answer the following questions. Use the table on page 207 for help.

22. Which metal will replace chromium but not aluminum from solution?

_____

23. Explain why silver will not react with hydrochloric acid but magnesium

will react with this acid. _____

_____

An electrochemical cell is set up as shown in Fig. 42-7. Use the diagram to complete items 24 – 29.

Fig. 42-7

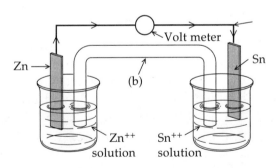

Which metal loses electrons? (24)_____ Which metal gains

electrons? (25)_____ What are the voltages of each

half-reaction? (26)_____ What is the voltage of the whole

cell? (27)_____ In which half-cell does oxidation take

place? (28)_____ In which half-cell does reduction take

place? (29)_____

Application of Electrochemistry

battery: two or more electrochemical cells connected together

corrosion: redox reactions that produce waste material from useful metals

electroplating: use of electric current to coat the surface of one metal with another metal

electrolysis: use of electricity to separate compounds into simpler substances

KEY IDEAS

Redox reactions in batteries produce useful electricity. Other redox reactions use electricity to refine metals or to plate one metal with another. Redox reactions also take place when metals rust or corrode.

Fig. 43-1

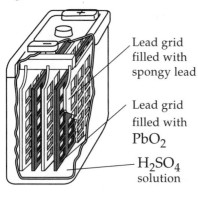

Lead grid filled with spongy lead

Lead grid filled with PbO_2

H_2SO_4 solution

Electricity is often used to plate costly silver or gold onto less expensive metals. This process, called electrolysis, is often used in the making of jewelry. A knowledge of redox reactions helps jewelry makers use electrolysis as a tool of their trade.

Inside Batteries. A **battery** (BAT-uhr-ee) is made of two or more electrochemical cells. The lead storage battery shown in Fig. 43-1 is used in a car or truck. Six 2-volt cells are joined together to make this 12-volt battery. In each cell, oxidation takes place at a lead plate. The electrons then flow to a lead dioxide grid where reduction takes place. Thus a redox reaction takes place.

A lead storage battery can be recharged. An outside source of electricity reverses the redox reaction process. The electrons are sent from the lead dioxide (PbO_2) back to the lead plate.

 1. Many electrochemical cells hooked together form a _____.

 2. Reversing the redox reaction will _____ a storage battery.

A flashlight battery such as the one shown in Fig. 43-2 is actually a single cell. The ions are in a paste of chemicals instead of in a watery liquid. For this reason, this type of battery is called a dry cell. All the parts of a dry cell are in a sealed case, so it can be easily carried around.

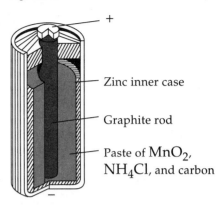

Fig. 43-2

+

Zinc inner case

Graphite rod

Paste of MnO_2, NH_4Cl, and carbon

−

Oxidation takes place at the outer container of a dry cell, which is made of zinc. Reduction occurs where the paste touches the graphite rod. In alkaline batteries, potassium hydroxide is used instead of ammonium chloride.

Redox Reactions that Waste Metals. Redox reactions that take place when metals oxidize cause **corrosion** (kuh-ROH-zhuhn). Corrosion wastes or destroys metal. Many metals oxidize to make a thin coating that almost stops further oxidation. Aluminum is such a metal. The dull haze of aluminum oxide can be seen on the outside of aluminum soda cans. This layer of aluminum oxide prevents water and oxygen from reaching the metal and causing more corrosion.

Rusting is the corrosion of iron. Water and oxygen are also needed to rust iron. In a complex reaction, iron loses electrons. The lost electrons then react with water and hydrogen. In a final reaction, rust is formed.

When rusting occurs, the water serves as a sort of salt bridge. The salt bridge is more effective if the salt NaCl is present in the water. This explains why cars rust faster in areas where large amounts of salt are used on roads.

Coatings of paint or of another metal are often used to cover iron and prevent corrosion. Steel bridges are painted to keep oxygen and water away from the metal. A tin can is really a steel can coated with a layer of protective tin. Galvanized iron is made by coating iron with a protective layer of zinc.

Some alloys, or mixtures of different metals, can also resist corrosion. Stainless steel is an alloy of iron that resists rusting. Stainless steel contains nickel and chromium as well as iron.

Products from Redox Reactions. During **electroplating** (ee-LEHK-troh-playt-ihng) an outside source of electricity moves one metal onto the surface of another metal. Study the diagram in Fig. 43-3. It shows silver being plated onto an iron spoon. A bar of silver (Ag) and an iron (Fe) spoon are hung in a solution of silver nitrate ($AgNO_3$). A source of electricity passes electrons through the iron and into the solution. Silver ions (Ag^+) gain electrons to form atoms of silver metal, which plate the iron spoon. More silver ions form from the bar of silver to replace those that plate the iron spoon.

Fig. 43-3

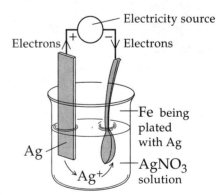

Electricity source

Electrons + − Electrons

Fe being plated with Ag

Ag

Ag^+

$AgNO_3$ solution

Tableware and jewelry are among the many products made by electroplating metals with silver or gold. Many computer chips are also electroplated with gold.

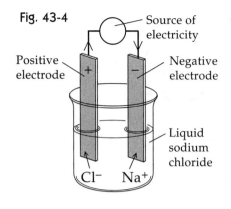

Fig. 43-4

Source of electricity

Positive electrode

Negative electrode

Liquid sodium chloride

Cl⁻   Na⁺

Metals found in nature usually exist as compounds. Some of these compounds are strongly bonded, requiring a lot of energy to force the elements apart. **Electrolysis** (ee-lehk-TRAHL-ih-sihs) uses an electric current to separate such compounds.

The electrolysis of sodium chloride (NaCl) is shown in Fig. 43-4. In this process, electrons flow from the source of electricity to the negative pole or electrode. Here sodium ions gain electrons and form sodium atoms. Chloride ions move to the positive pole, or electrode. Here they lose electrons and form chlorine gas.

Chlorine gas produced by electrolysis is used to purify water, to bleach materials, and to make many plastics. Electrolysis also produces other useful elements, such as aluminum. This metal is used to make cans, foil wrap, home siding, and many other products.

The difference between a cell that makes electricity and a cell that uses electricity is shown in Fig. 43-5.

Fig. 43-5

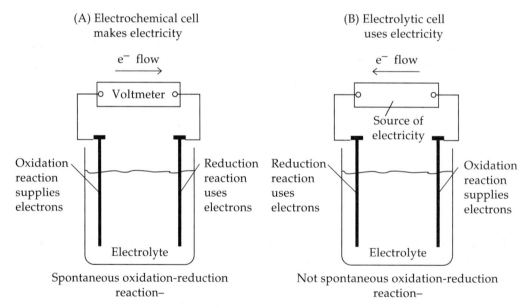

(A) Electrochemical cell makes electricity

e⁻ flow

Voltmeter

Oxidation reaction supplies electrons

Reduction reaction uses electrons

Electrolyte

Spontaneous oxidation-reduction reaction–

(B) Electrolytic cell uses electricity

e⁻ flow

Source of electricity

Reduction reaction uses electrons

Oxidation reaction supplies electrons

Electrolyte

Not spontaneous oxidation-reduction reaction–

Check Your Understanding

Fill in the blanks.

3. Two or more electrochemical cells joined together form a

    _____.

4. In a 12-volt lead storage battery, oxidation takes place at a

    _____.

5. Reduction in the storage battery takes place at the _____.

6. Each cell in a storage battery makes _____ volts of electricity.

7. Storage batteries are recharged by an outside source of

_____.

8. A flashlight battery is really a _____.

List three ways to prevent corrosion.

9. _____

10. _____

11. _____

12. An outside source of _____ is used in both electroplating and electrolysis.

Underline the correct word or phrase in items 13-17.

13. In a lead storage battery, **(a)** acid is poured on lead. **(b)** two or more cells are joined together. **(c)** only oxidation takes place. **(d)** only reduction takes place.

14. The outer container of a dry cell is made of **(a)** ammonium chloride. **(b)** graphite. **(c)** potassium hydroxide. **(d)** zinc.

15. A metal that resists corrosion by making a thin oxide coating is **(a)** aluminum. **(b)** iron. **(c)** oxygen. **(d)** salt.

16. A process used to separate a metal from the rest of a compound is **(a)** oxidation. **(b)** corrosion. **(c)** electroplating. **(d)** electrolysis.

17. What happens at the positive electrode during the electrolysis of liquid sodium chloride? **(a)** Sodium atoms gain electrons. **(b)** Sodium ions gain electrons. **(c)** Chlorine atoms lose electrons. **(d)** Chloride ions lose electrons.

Write a sentence to answer question 18.

18. In the presence of oxygen and water, metals oxidize or corrode. Explain why aluminum airplanes do not decompose during a rain storm.

_____

_____

Summary

- The loss of electrons is oxidation. The gain of electrons is reduction. The oxidizing agent gains electrons, becomes more negative, and is reduced. The reducing agent loses electrons, becomes more positive, and is oxidized.

- Oxidation numbers show the number of electrons gained, lost, or shared during bonding. A set of rules is used to find the oxidation numbers of the atoms in elements and compounds. The sum of all the oxidation numbers in a substance is always zero. The term redox means oxidation-reduction reaction. During redox reactions, oxidation numbers change.

- An electrochemical cell is made of:
  two half-cells—one for oxidation and one for reduction;
  a salt bridge that carries ions between the half-cells; and
  an external conductor that carries electrons between the half-cells.

- A table of standard reduction voltages is used to calculate the voltages of cells and to predict spontaneous reactions. A spontaneous reaction takes place without the addition of outside energy.

- Batteries are electrochemical cells joined together. A flashlight battery is really a dry cell.

- During corrosion, metals form waste products. Electroplating uses electricity to cover one metal with another metal. Electrolysis uses electricity to separate the elements making up metallic compounds.

For Your Portfolio

1. Imagine you are a newscaster reporting on electric cars. Prepare an informational videotape about the batteries used in electric automobiles.

2. Explain to a classmate why iron oxidizes quickly to form rust but aluminum does not.

3. Imagine that a toy needs a battery that has a spherical shape. Design such a dry cell.

4. You are to help teach the electroplating process to a class of third grade students. Design a dance to demonstrate the plating process.

5. Corrosion of iron wastes almost one-fourth of the iron that is refined annually. Make a concept map that shows how we waste this resource and how conservation of this resource might be increased.

1. Which of the following always takes place in a redox reaction?

    (a) oxidation only (b) both oxidation and reduction (c) reduction only
    (d) neutralization

2. In a redox reaction, the oxidizing agent

    (a) is reduced (b) loses electrons (c) gains protons (d) is oxidized

3. What is the oxidation number of sulfur in H_2SO_3?

    (a) –2 (b) 0 (c) +6 (d) +4

4. If a redox reaction is forced to take place by using electricity, the process is called

    (a) ionization (b) electrolysis (c) spontaneous (d) voltage

5. In the reaction $Al + Cr^{3+} \rightarrow Cr + Al^{3+}$, the oxidizing agent is

    (a) Cr^{3+} (b) Al (c) Cr (d) Al^{3+}

6. In the reaction $Mg + 2H^+ \rightarrow H_2 + Mg^{2+}$, the reducing agent is

    (a) Mg (b) $2H^+$ (c) H_2 (d) Mg^{2+}

7. During the electrolysis of sodium chloride (NaCl), which reaction takes place at the negative electrode?

    (a) $Cl_2^0 + 2e^- \rightarrow 2Cl^-$ (b) $Na^0 - e^- \rightarrow Na^+$
    (c) $Na^+ + e^- \rightarrow Na^0$ (d) $2Cl^- - 2e^- \rightarrow Cl_2^0$

8. Which of the following is a redox reaction?

    (a) $Na^+ + Cl^- \rightarrow NaCl$ (b) $NaOH + HCl \rightarrow HOH + NaCl$
    (c) $BaCl_2 + Na_2SO_4 \rightarrow BaSO_4 + 2NaCl$ (d) $2Na + Cl_2 \rightarrow 2NaCl$

9. In an electrochemical cell, ions move from one half-cell to another half-cell through

    (a) electrodes. (b) an external conductor. (c) a salt bridge. (d) a battery.

10. In which of the following compounds is the oxidation number of nitrogen the lowest?

    (a) N_2O (b) NO_2 (c) NH_3 (d) N_2

11. In the reaction $2H_2O_2 \rightarrow 2H_2O + O_2$, the oxidation number of oxygen

    (a) increases only. (b) decreases only. (c) remains the same.
    (d) both increases and decreases.

Use the table on page 207 to help answer the following. Circle your answer.

12. What is the E^o of the half reaction $Ag^+ + e^- \rightarrow Ag(s)$?

    (a) +0.20   **(b)** –0.20   **(c)** +0.80   **(d)** –0.80

13. Which of the following metals will replace nickel in a spontaneous reaction?

    (a) Ag   **(b)** Au   **(c)** Cu   **(d)** Cr

14. Which of the following metals will react with HCl?
    (a) Ag   **(b)** Mg   **(c)** Cu   **(d)** Au

15. The reaction in an electrochemical cell is: $2Ag^+ + Cu \rightarrow Cu^{2+} + 2Ag$. What is the E^o for this cell?

    (a) +0.46 v   **(b)** +0.80 v   **(c)** +1.14 v   **(d)** +0.34 v

Answer one of the following questions.

16. Describe what happens during the processes of electrolysis and electroplating.

17. Explain why if oxidation takes place in a chemical reaction, then reduction must also take place.

Organic Chemistry

Do you recycle materials made of plastic? Many communities now require recycling of plastics such as those used in soda bottles and milk jugs. Plastics are made of compounds known as polymers. Polymers are used in so many products that our lives would be quite different without them. Next time you are in a supermarket, notice how many items are in wrappers or containers made of plastic. You will notice that most meats are packaged in plastic. And rolls of plastic bags made to hold fruits and vegetable are found in the produce section. See what items you can find that are not packaged in plastic.

In addition to their use in plastics, polymers have many other uses. They are used to make cloth (nylon and polyester) and waterproof coatings. Because they are good insulators, polymers are used to coat electrical wires. Polymers are strong and can easily be molded into different shapes (that's why they make good containers). However, because polymers do not break down, plastic containers and other materials made of polymers create a disposal problem.

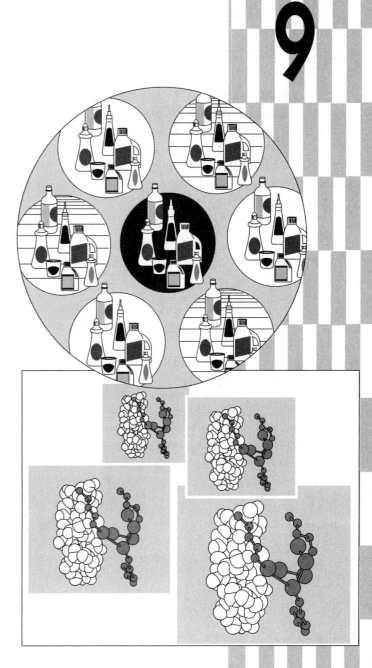

Hydrocarbons 1

Key Words

organic compound: compound containing carbon atoms

hydrocarbon: organic compound with only carbon and hydrogen atoms

alkane: hydrocarbon with only single bonds between neighboring carbon atoms

alkene: hydrocarbon with a double bond between two neighboring carbon atoms

alkyne: hydrocarbon with a triple bond between two neighboring carbon atoms

KEY IDEAS

Hydrocarbons are compounds that contain only carbon and hydrogen atoms. These atoms join together in different ways to form many useful compounds.

Some petroleum technicians work to discover large underground pockets of petroleum (oil) and natural gas. Other workers supply these substances in a useful form. Both oil and natural gas are hydrocarbons burned to produce energy.

Carbon: A Special Element. Almost any compound that contains carbon is known as an **organic compound** (awr-GAN-ihk KAHM-pownd). Any organic compound formed only of carbon and hydrogen atoms is a **hydrocarbon** (hy-dro-KAHR-buhn). Most fuels used in homes and industry are hydrocarbons.

Natural gas, or methane, is a fuel burned in many homes to provide heat and hot water. Methane is the simplest of all hydrocarbons. In methane (CH_4), one carbon atom is joined with four hydrogen atoms. Look at Fig. 44-1. Notice that a carbon atom has four electrons in its outer energy level. Hydrogen atoms have only one electron. In methane, each carbon electron forms a pair, or bond, with one hydrogen electron. This is shown in Fig. 44-2. In organic compounds, chemists usually use a straight line instead of dots to show a bond between the different atoms. See Fig. 44-3.

Fig. 44-1   Fig. 44-2 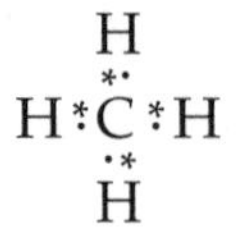   Fig. 44-3

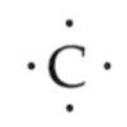

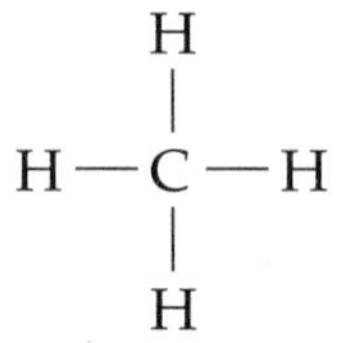

$$\cdot \overset{\displaystyle\cdot}{\underset{\displaystyle\cdot}{C}} \cdot$$

$$\overset{\textstyle H}{\underset{\textstyle H}{H\!:\!C\!:\!H}}$$

$$\overset{\textstyle H}{\underset{\textstyle H}{\underset{\textstyle |}{\overset{\textstyle |}{H\!-\!C\!-\!H}}}}$$

A carbon atom has four electrons to share. It bonds easily with hydrogen atoms, each of which has one electron to share. This is one reason why there are so many different hydrocarbons.

Carbon. Carbon's special bonding property allows carbon atoms to link up with each other. Sometimes, carbon atoms form very long chains. However, chains of carbon atoms are not all alike. For example, **alkanes** (AL-kaynz) are hydrocarbons in which the carbon atoms are joined to each other by one pair of electrons. An alkane with two carbon atoms joined by a single bond is shown in Fig. 44-4. Two neighboring carbon atoms in a chain sometimes have a double bond. A double bond forms when two pairs of electrons are shared. **Alkenes** (AL keenz) are hydrocarbons in which two neighboring carbon atoms share two pairs of electrons. Fig. 44-5 shows an alkene. Gasolines with greater amounts of alkanes than alkenes burn more efficiently.

Hydrocarbons may also have triple bonds. In **alkynes,** (AL-kynz) two neighboring carbon atoms share three pairs of electrons, forming a triple bond. The simplest alkyne, acetylene, is shown in Fig. 44-6. Welders use acetylene gas in blowtorches to cut and shape metals.

Fig. 44-4

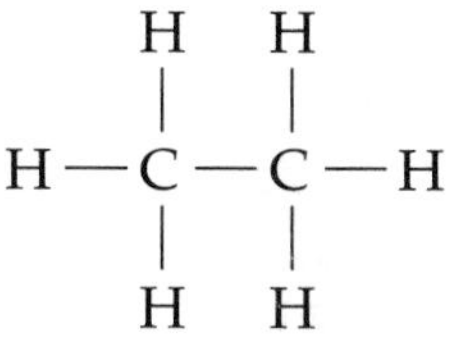

$$H-\overset{\displaystyle H}{\underset{\displaystyle H}{C}}-\overset{\displaystyle H}{\underset{\displaystyle H}{C}}-H$$

Fig. 44-5

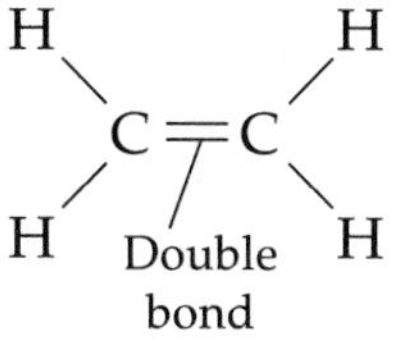

$$\overset{H}{\diagdown}\underset{H}{\diagup}C\!=\!C\overset{\diagup H}{\diagdown H}$$

Double bond

Fig. 44-6

$$H-C\!\equiv\!C-H$$

Triple bond

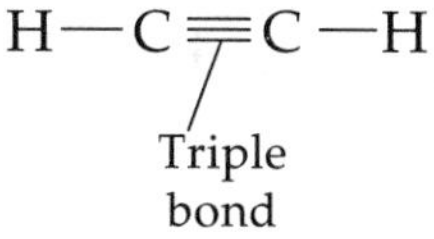

  1. **What is the difference between an alkene and an alkane?**

_____

_____

_____

Sometimes the carbons in a hydrocarbon bond together to form a ring instead of a chain. Benzene is an example of a ring hydrocarbon. The ring has six carbon atoms, each of which bonds to a hydrogen atom. A benzene ring has three double bonds. See Fig. 44-7. The three double bonds move back and forth between neighboring carbon atoms.

Fig. 44-7

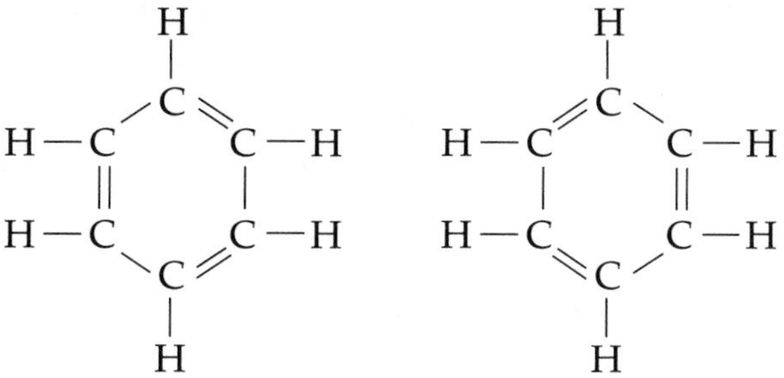

Fig. 44-8 shows the relationship among different types of organic compounds.

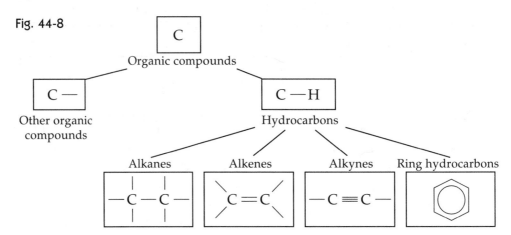

Fig. 44-8

Check Your Understanding

Write a sentence explaining the connection between each pair of terms.

2.   organic compound, carbon _____

_____

3.   hydrocarbon, organic compound _____

_____

4.   alkane, hydrocarbon _____

_____

5.   alkene, double bond _____

_____

6.   alkyne, neighboring carbon atoms _____

_____

Complete each electron dot diagram.

7.   hydrogen
     atom
            H

8.   methane

            ·Ċ·

9.   alkane
                H   H

            H   C   C   H

                H   H

10.   alkyne

            H   C   C   H

11. Explain the difference between any organic compound and a hydrocarbon.

_____

_____

12. With an electron dot diagram, show the outer electron energy level of a carbon atom.

13. How many carbon-hydrogen bonds are present in a molecule of methane? _____

14. Draw an alkane that has three carbon atoms. Include the bonds and the hydrogen atoms around each carbon.

15. Draw an example of an alkene and an alkyne, each containing three carbon atoms.

16. What is a triple bond? _____

_____

17. Suppose an alkane and an alkyne have five carbon atoms each. Which compound would have more hydrogen atoms? Explain.

_____

_____

_____

Lesson 45 Hydrocarbons 2

KEY IDEAS

Some hydrocarbons contain functional groups, or groups of atoms other than just carbon and hydrogen. These groups give the compounds different properties. As a result, the compounds can be used for many purposes.

Dietitians plan good, healthful meals for people. Some people must limit the sugar in their diets. So dietitians often include sugar substitutes in their meals. Both sugar and sugar substitutes are hydrocarbons containing groups of atoms other than carbon and hydrogen.

Hydrocarbons with Oxygen. Many hydrocarbons contain groups of atoms besides carbon and hydrogen. Oxygen atoms are often present in these groups. A **functional group** (FUNK-shuh-nuhl) is a group of atoms that give hydrocarbons certain properties. These compounds can then be used for different purposes.

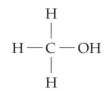

Fig. 45-1

An **alcohol** (AL-kuh-hawl) is a type of hydrocarbon that contains the functional group -OH (oxygen and hydrogen). In an alcohol, a hydrogen atom is replaced by an -OH functional group. The oxygen of the -OH group bonds with a carbon atom. The structure of methanol, the simplest alcohol, is shown in Fig. 45-1. Methanol burns easily to produce energy. In fact, car builders are trying to find ways to use methanol as fuel for car engines.

Fig. 45-2

Other useful substances are made from alcohols, too. For example, ethylene glycol is an alcohol with two -OH groups. See Fig. 45-2. Car mechanics know that ethylene glycol makes a good antifreeze in car engines. Why? Because of its -OH groups, ethylene glycol dissolves completely in water. In addition, a solution of ethylene glycol and water in a car radiator will not freeze easily in cold weather or boil over in hot weather.

Hydrocarbons with the -CHO functional group at the end of the carbon chain are known as **aldehydes** (AL-duh-hyedz). Carbon and oxygen share a double bond in aldehydes. Fig. 45-3 shows the structure of the simplest aldehyde, methanal. The common name for methanal is formaldehyde. Perhaps you have smelled this liquid in the biology lab, where it is often used to preserve animal specimens. In the laboratory, methanal is produced from the alcohol methanol. This aldehyde is also important in making plastics.

Fig. 45-3

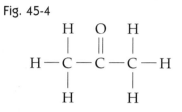

The functional group C=O is present in all **ketones**. Figure 45-4 shows how a carbon atom in the middle of a hydrocarbon chain can be joined by a double bond to an oxygen atom. Notice that the -OH group does not appear at the end of the chain. Propanone, also called acetone, is a ketone known to anyone who must remove nail polish. It also dissolves paints and plastics.

Fig. 45-4

 1. How do the functional groups of aldehydes and ketones differ?

_____

_____

Hydrocarbons containing the functional group -COOH are called **organic acids**. The functional group -COOH includes a double bond between the carbon and one oxygen. Most dietitians tell their patients to eat lots of foods with citric acid. Citric acid, with three -COOH groups, is found in citrus fruits such as oranges and grapefruits.

If you have had surgery, you may have been given **ether** (EE-thuhr) as an anesthetic. Ethers contain the functional group COC. The structure of diethyl ether, the ether used in surgery, is shown in Fig. 45-5. As in all other ethers, the COC group appears in the center of the hydrocarbon chain.

Fig. 45-5

$$H-\overset{\overset{\displaystyle H}{|}}{\underset{\underset{\displaystyle H}{|}}{C}}-\overset{\overset{\displaystyle H}{|}}{\underset{\underset{\displaystyle H}{|}}{C}}-O-\overset{\overset{\displaystyle H}{|}}{\underset{\underset{\displaystyle H}{|}}{C}}-\overset{\overset{\displaystyle H}{|}}{\underset{\underset{\displaystyle H}{|}}{C}}-H$$

It is important to study the makeup of organic compounds because the presence of certain atoms results in different properties. For example, think about table sugar and the artificial sweetener known as NutraSweet™. Both have many functional groups. The functional groups affect their taste and the way they are used by the body. So artificial sweeteners can be used in very small amounts to make foods taste good, with hardly any extra calories. Some scientists think the benzene ring of the artificial sweetener is responsible for its sweetness.

TAKE ANOTHER LOOK

Figure 45-6 shows the relationships among hydrocarbons and hydrocarbon derivatives.

Fig. 45-6

```
                    ┌─────────┐
                    │  C — H  │
                    └─────────┘
                   Hydrocarbons
```

— OH	— CHO	— C = O	— COOH	COC
Alcohols	Aldehydes	Ketones	Organic acids	Ethers

Check Your Understanding

In the blank beneath each diagram, write whether the compound shown is a ketone, an organic acid, an aldehyde, an ether, or an alcohol.

2.
```
        OH
        |
   H — C — H
        |
        H
```
_____

3.
```
        H    O
        |   //
   H — C — C
        |   \
        H    H
```
_____

4.
```
        H   O   H
        |   ||  |
   H — C — C — C — H
        |       |
        H       H
```
_____

5.
```
        H
        |     O
   H — C — C //
        |     \
        H      OH
```
_____

6.
```
        H       H
        |       |
   H — C — O — C — H
        |       |
        H       H
```
_____

Complete each of the four diagrams below.

7. alcohol
```
     H   H
     |   |   |
   — C — C — C   OH
     |   |   |
     H   |   H
       — C —
         |
```

8. aldehyde
```
           H        O
           |    |
      H    C — C    C
           |    |
                      H
```

9. organic acid
```
        H
        |    |
   H — C — C   C //
        |         \
        H
```

10. methylethyl ether
```
        H              H
        |              |
   H — C —      — C —  | — H
        |              |
        H              H
```

226   Unit 9   Organic Chemistry

11. What is a functional group? _____
_____

12. List two properties of the alcohol ethylene glycol.

_____
_____

13. Draw a picture of the functional group of an organic acid in the space below.

14. Draw the structure of an organic compound with four carbon atoms and two alcohol groups.

15. What functional group is found in aldehydes? What kinds of bonds does this functional group have? _____

_____
_____

16. Draw a five-carbon compound that contains one ketone group.

17. How could you change the diagram you made to show an ether?

_____
_____

Lesson 46
Reactions of Organic Compounds

Key Words

substitution reaction:	reaction in which atoms or functional groups are substituted for one or more hydrogen atoms in an alkane
addition reaction:	reaction in which one or more atoms or functional groups are added to an alkene or alkyne
dehydration reaction:	reaction in which two substances are combined by the removal of water
polymer:	large molecule made of simple molecular units repeated many times
polymerization reaction:	a reaction that produces a polymer from monomers

KEY IDEAS

Many organic reactions involve the union of a functional group with a hydrocarbon. These reactions usually occur by oxidation, addition, substitution, or dehydration. Polymers form in a different way. In polymerization, two or more small molecules, or monomers, combine to form a larger molecule called a polymer.

Perhaps you have a jacket made of synthetic fleece. The fleece could actually be a polymer made from recycled plastic soda bottles. As a laboratory technician, you could have been involved in many steps in its production. Or you may be interested in synthetic dyes used to produce designs for fabrics. Many fabrics and their dyes are synthetic polymers.

Organic Reactions. If your home is heated with gas, it is probably heated with methane or natural gas. When methane (or another fuel) burns, it combines with oxygen to produce carbon dioxide and water as shown in the following equation. Recall that a reaction in which oxygen combines with another substance is an *oxidation reaction*.

$$CH_4 + 2O_2 \longrightarrow CO_2 + 2H_2O$$

Suppose you want to combine chlorine with methane. Each carbon atom in the methane is already bonded to a hydrogen atom by a single bond. So you must use a reaction that substitutes an atom of chlorine for an atom of hydrogen. Actually, you can produce four compounds in which you replace one, two, three, or all four of the hydrogen atoms with chlorine atoms. See Fig. 46-1. A **substitution reaction** has taken place to produce each compound.

Fig. 46-1

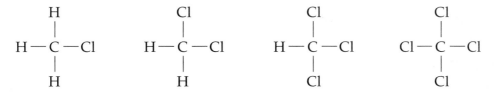

Now suppose you want to combine bromine with ethene. Ethene has a double bond. To combine these substances, you can break the double bond and add two bromine atoms to ethene. See Fig. 46-2. An **addition reaction** has taken place.

Fig. 46-2

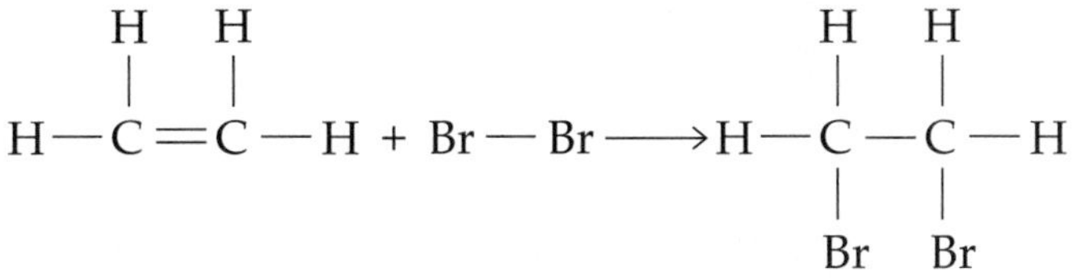

$$H-C=C-H + Br-Br \longrightarrow H-C-C-H$$

Ethene   Bromine

Now suppose you want to produce an alcohol. In any alcohol, one or more hydroxyl (-OH) groups are substituted for one or more hydrogen atoms of a hydrocarbon.

Alcohols have many uses of their own. They are also important because aldehydes, ketones, organic acids, and ethers are produced from alcohols. Aldehydes, ketones, and organic acids are produced by oxidation reactions.

The simplest aldehyde, formaldehyde, can be produced by an oxidation reaction. In this reaction, methyl alcohol reacts with an oxidizing agent. In the equation shown in Fig. 46-3, the oxidizing agent is shown as [O].

Fig. 46-3

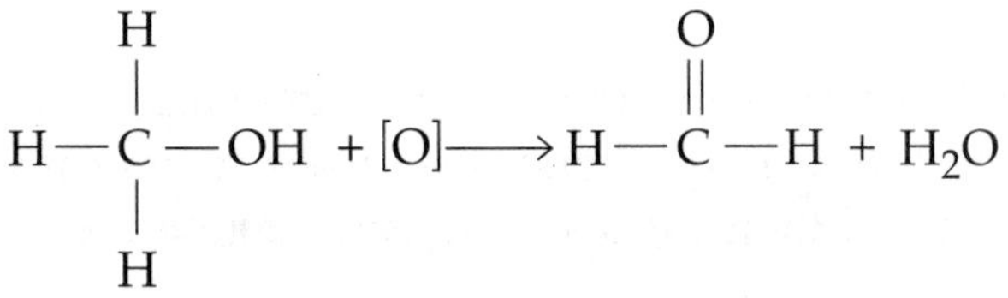

$$H-C-OH + [O] \longrightarrow H-C-H + H_2O$$

Methyl alcohol   Formaldehyde

Ethers are produced by dehydration. The equation in Fig. 46-4 shows the **dehydration reaction** (dee-hy-DRAY-shuhn) that produces diethyl ether. In this reaction, ethyl functional groups from two molecules of ethyl alcohol become bonded to a single oxygen atom. The result of the reaction is an ether.

Fig. 46-4

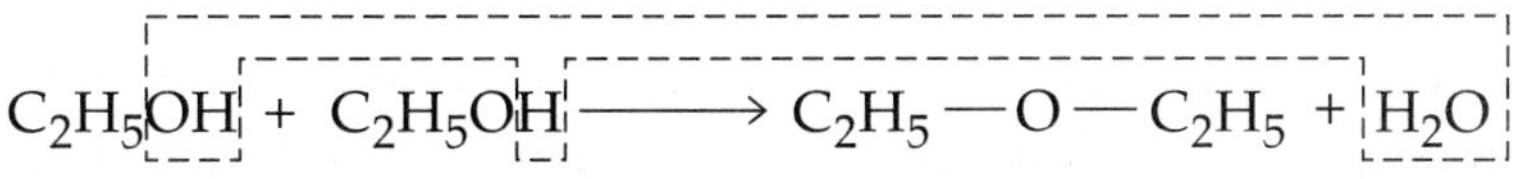

$$C_2H_5OH + C_2H_5OH \longrightarrow C_2H_5-O-C_2H_5 + H_2O$$

Ethyl alcohol   Ethyl alcohol   Diethyl ether

1. How does an oxidation reaction differ from a dehydration reaction?

_____

_____

_____

Polymers. **Polymers** (pahl-uh-muhrz) are large molecules made of small molecules that are joined together. Synthetic polymers are used to produce plastics, fibers, and rubbers. In polymerization, monomers, or small organic molecules, are joined together to form large molecules. In a **polymerization reaction** (pahl-uh-muhr-uh-ZAY-shuhn), the monomers may be the same compound or different compounds.

In nature, only one kind of monomer is joined together to form a polymer. In the laboratory, chemists can combine two different kinds of monomers. The polymer produced is called a copolymer. Copolymers have made thousands of synthetic polymers possible.

Fig. 46-5 shows the types of compounds formed by the different reactions of hydrocarbons. In this diagram, the letter **R** stands for a group of carbon and hydrogen atoms.

Fig. 46-5

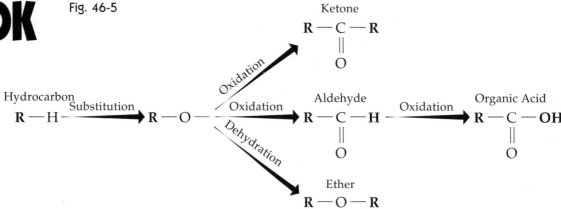

Check Your Understanding

Use Fig. 46-5 to fill in the blanks in the sentences below.

The organic reactions, in order, needed to derive an organic acid from a simple hydrocarbon are **(2)**_____, oxidation, **(3)** _____. A compound produced by dehydration has had **(4)**_____ removed.

The three types of organic compounds derived by the oxidation of alcohols are **(5)**_____, **(6)**_____, and **(7)** _____ . Dehydration of alcohols results in the formation of **(8)**_____. As the hydrocarbon at the left (R-H) is an alkane, it has only a (9)_____ bond. It has no double or

(10) _____ bonds. Therefore, all the other organic

compounds are formed by (11)_____ reactions, not by

addition reactions. An alkane can be combined with chlorine or bromine

in (12) _____ reactions.

13. Complete the equation for a substitution reaction between the alkane methane and a molecule of bromine.

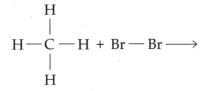

$$\begin{array}{c} H \\ | \\ H-C-H \\ | \\ H \end{array} + Br-Br \longrightarrow$$

14. Complete the equation for an addition reaction between the simplest alkene and a molecule of chlorine.

$$\begin{array}{cc} H & H \\ | & | \\ H-C=C-H \end{array} + Cl-Cl \longrightarrow$$

15. In nylon, two monomers combine by the removal of water. What type of reaction takes place?

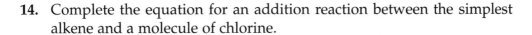

16. What is a copolymer?

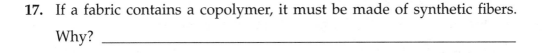

17. If a fabric contains a copolymer, it must be made of synthetic fibers. Why? _____

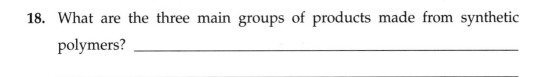

18. What are the three main groups of products made from synthetic polymers?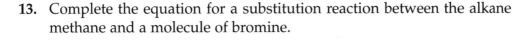

_____

Biochemistry

carbohydrates:	aldehydes or ketones with many -OH groups attached
polysaccharide:	long chain of simple sugars
lipids:	group of compounds including fats, oils, and waxes
triglyceride:	animal fat composed of fatty acids and glycerol
proteins:	group of compounds made of long chains of amino acids
amino acids:	organic acids that contain a carbon-nitrogen bond
peptide:	compound formed by the bonding together of amino acids
enzyme:	protein that speeds up a chemical reaction
nucleic acids:	compounds composed of long chains of sugars, phosphates, and nitrogen bases

KEY IDEAS

Biochemistry is the chemistry of living things. The reactions that take place within living things rely on certain organic compounds. Reactions that involve these compounds control the growth and regulation of living cells.

Nutritionists plan different diets for people in different situations. For healthy people, they suggest diets that will help maintain good health. They may also suggest foods that avoid weight gain. For people who have certain health problems, nutritionists suggest diets to improve health. In order to know what foods to include or avoid in certain diets, nutritionists must understand body chemistry.

Sugars and Starches. **Carbohydrates** (kahr-boh-HY-drayts) are aldehydes or ketones with many -OH groups attached. Carbohydrates may be simple or complex molecules. The simplest carbohydrates are simple sugars.

If water is removed from two simple sugars, they combine to form a double sugar. Common table sugar is a double sugar. The reaction forming table sugar is shown in Fig. 47-1.

Fig. 47-1

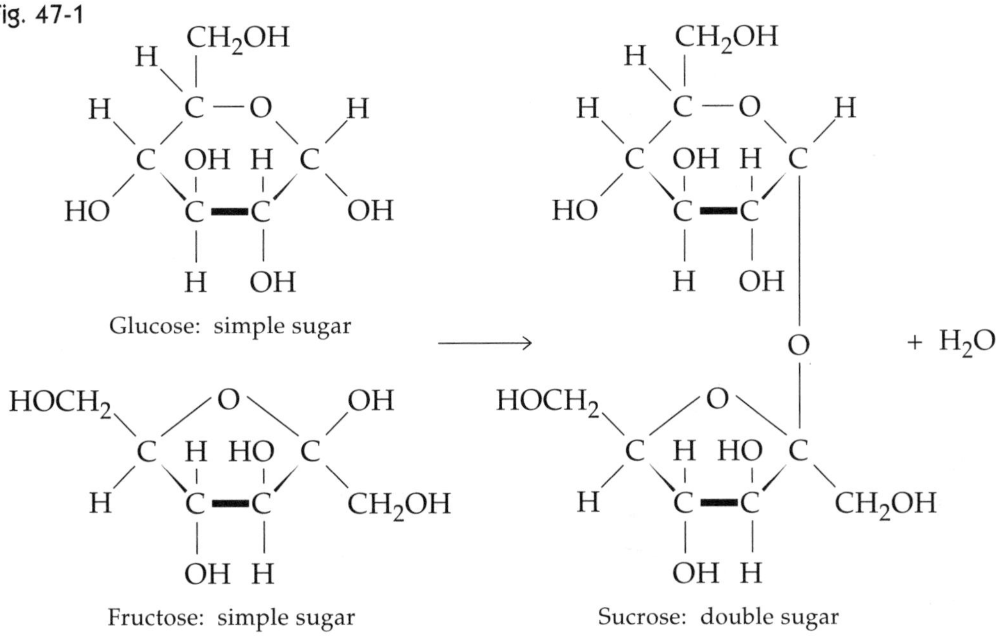

Glucose: simple sugar

$\longrightarrow$

$+ H_2O$

Fructose: simple sugar   Sucrose: double sugar

When simple sugars continue to bond together, a long chain is formed. Such a chain of simple sugars is called a **polysaccharide** (pahl-ih-SAK-uh-ryed). Starches are polysaccharides formed from long chains of the sugar glucose. Glucose is stored in plants and animals as starch. This starch can then be broken down into sugar as it is needed. Sugar molecules enter into reactions that provide energy for cells.

Plant cell walls are made of another polysaccharide called cellulose. This is a very sturdy compound that gives structure to many plants.

Fats and Oils. Fats, oils, and waxes make up a group of compounds called **lipids**. Lipids do not dissolve well in water. You can see this if you try to mix water and cooking oil. No matter how much you shake or stir them together, they keep separating. They do not go into solution.

Animal fats and oils are a type of lipid called a **triglyceride** (try-GLIHS-uhr-eyed). A triglyceride is a compound formed by combining fatty acids and glycerol. Fatty acids are organic acids. Glycerol is an alcohol. The general reaction that forms a triglyceride is shown in Fig. 47-2. In this reaction, water is a product. In the human body, fats are stored as triglycerides.

Fig. 47-2

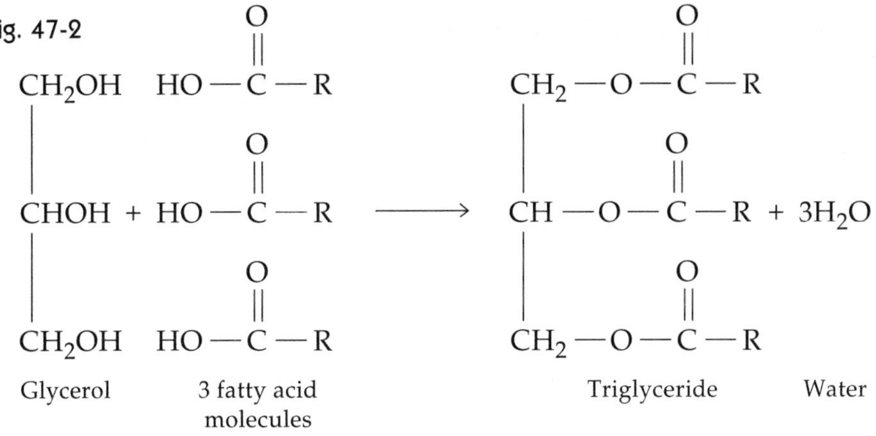

Glycerol   3 fatty acid   Triglyceride   Water
                molecules

Fig. 47-3

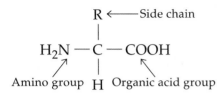

R ←——Side chain

$H_2N — C — COOH$

Amino group   H   Organic acid group

Proteins. Many structures of living things are made of proteins. **Proteins** (PROH-teenz) are long chains of compounds called **amino acids** (uh-MEE-noh). The general structure of an amino acid is shown in Fig. 47-3.

When amino acids bond together, they form a compound called a **peptide** (PEHP-tyed). The bond formed between them is called a peptide bond. Fig. 47-4 shows how this bond forms. Notice that water is a product of this reaction.

Fig. 47-4

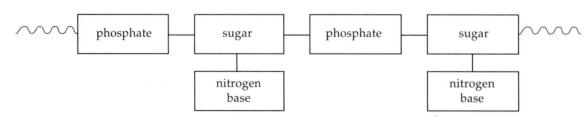

Amino acid   Amino acid   Peptide

Amino acids can join together over and over again. This addition makes the chain longer. Proteins are chains that have 100 or more amino acids.

Many reactions that take place inside living cells involve different proteins. One type of protein that speeds a chemical reaction is called an **enzyme** (EHN-zyem). The enzyme is not changed by the reaction. Many different enzymes are present in living things.

 1. **How are amino acids, peptides, and proteins related?**

_____

_____

Nucleic Acids. One very important compound in living things is DNA. DNA makes up genes. Genes pass traits from generation to generation. They do this by controlling the formation of proteins.

DNA belongs to a group of chemicals called **nucleic acids** (noo-KLEE-ihk). These are made up of chains of smaller compounds. The smaller compounds that form nucleic acids contain a phosphate group, a simple sugar, and a nitrogen base. These compounds continue to bond together in the order shown in Fig. 47-5.

Fig. 47-5

phosphate	sugar	phosphate	sugar
	nitrogen base		nitrogen base

There are four different nitrogen bases in DNA. These bases pair with each other in a certain way. Because of this, the DNA molecule can duplicate itself. Another nucleic acid, RNA, helps this process.

DNA chains are very long. The number of different combinations of the four bases is huge. So DNA patterns are different from person to person. This means a person's DNA pattern can be like a fingerprint. Those working in police science can use DNA "prints" to identify people.

Fig. 47-6 shows the groups of biochemicals that are involved in the growth and regulation of cells.

Fig. 47-6

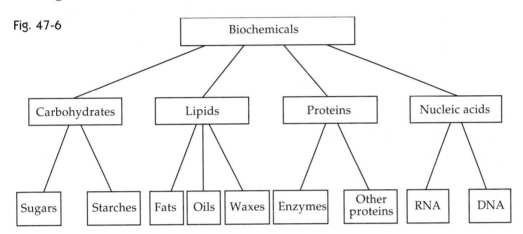

Check Your Understanding

Write a sentence explaining the connection between each pair of words.

2. carbohydrate, sugar _____

_____

3. polysaccharide, starch _____

_____

4. lipid, triglyceride _____

_____

5. protein, amino acid _____

_____

6. nucleic acid, DNA _____

_____

Several different reactions are shown below. On each numbered line identify what class of compound can be formed by the reaction shown. Choose your answers from the following: protein, carbohydrate, lipid, nucleic acid.

7. The reaction below produces a _____

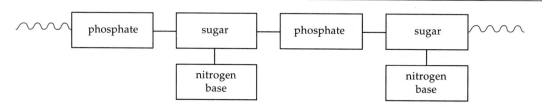

8. The reaction below produces a _____

$$CH_2OH \quad HO-\overset{\overset{O}{\|}}{C}-R$$
$$CHOH + HO-\overset{\overset{O}{\|}}{C}-R \longrightarrow$$
$$CH_2OH \quad HO-\overset{\overset{O}{\|}}{C}-R$$

Glycerol   3 fatty acid molecules

$$CH_2-O-\overset{\overset{O}{\|}}{C}-R$$
$$CH-O-\overset{\overset{O}{\|}}{C}-R + 3H_2O$$
$$CH_2-O-\overset{\overset{O}{\|}}{C}-R$$

Triglyceride   Water

9. The reaction below produces a _____

$$H_2N-\overset{\overset{R}{|}}{\underset{H}{C}}-\overset{\overset{O}{\|}}{C}-\boxed{OH + H}-N-\overset{\overset{R}{|}}{\underset{H}{C}}-\overset{\overset{O}{\|}}{C}-OH \longrightarrow H_2N-\overset{\overset{R}{|}}{\underset{H}{C}}-\overset{\overset{O}{\|}}{C}-N-\overset{\overset{R}{|}}{\underset{H}{C}}-\overset{\overset{O}{\|}}{C}-OH + H_2O$$

10. The chemicals below form a _____

phosphate — sugar — phosphate — sugar

sugar → nitrogen base

sugar → nitrogen base

11. What three elements make up a simple sugar? _____

_____

12. Simple sugars are the simplest form of carbohydrates. Carbohydrates
are _____ or _____, which have many
-OH groups attached.

13. In your own words, explain the reaction that forms common table
sugar. _____

_____

_____

14. What is a polysaccharide? _____

_____

15. What type of compound makes up the group known as lipids?

_____

16. In the human body, what compound stores fat?

_____

17. List two facts you have learned about enzymes.

_____

_____

18. What is a peptide bond? _____

_____

19. Can peptide bonds be found in proteins? _____ In
carbohydrates? _____ In lipids? _____Why?

_____

_____

20. Why are nucleic acids important to living things?

_____

_____

Summary

- Hydrocarbons are compounds that contain only carbon and hydrogen atoms. These atoms can be joined in many different ways resulting in a large number of different compounds.

- Alkanes, alkenes, and alkynes are different types of hydrocarbons. These compounds differ from each other in the types of bonds they form. Alkanes have only single bonds between carbon atoms; alkenes have at least one double bond, and alkynes have at least one triple bond.

- Benzene is a ring hydrocarbon. In a benzene ring, the double bonds move back-and-forth between neighboring carbon atoms.

- Sometimes hydrocarbons contain functional groups. Different functional groups give hydrocarbons different properties. Alcohols contain an -OH functional group; aldehydes contain the functional group -CHO; ketones contain the functional group -C=O; and organic acids contain the functional group -COOH.

- Organic compounds usually undergo one of four types of reactions: oxidation, addition, substitution, or dehydration.

- In an oxidation reaction, oxygen combines with another substance. In a substitution reaction, one chemical substance is substituted for another. In an addition reaction, a substance is added to a compound, usually by breaking double bonds. In a dehydration reaction, two substances are combined by the removal of water.

- Carbohydrates have many -OH groups attached to carbon atoms. Carbohydrates may be simple or complex. The simplest carbohydrates are simple sugars.

- Polysaccharides are formed from long chains of simple sugars. If a polysaccharide is formed from long chains of glucose, it is called a starch. Starches can be broken down as needed for energy or they can provide structural support for plants.

- Fats, oils, and waxes are called lipids. Animal fats are a special type of lipid called triglycerides.

- Much of the growth and regulation in living things is controlled by proteins. These are made of long chains of compounds called amino acids that are held together by peptide bonds. One type of protein, called an enzyme, speeds chemical reactions.

- The genetic material of living things is made up of compounds called nucleic acids. These are referred to as RNA and DNA. Nucleic acids control cell growth and regulation by directing the synthesis of proteins.

For Your Portfolio

1. Construct a model of an alkane, an alkene, and an alkyne. You may use any materials you like (clay, toothpicks, jellybeans, and so on). Be sure to clearly represent the number of bonds.

2. Some gasolines claim to reduce engine "knock" Investigate the problem of engine knock and find out what is done to gasoline to reduce this problem. Explain the process in a poster.

3. Oil spills are a large problem for living things. When oil is spilled in water, it can be carried great distances thus spreading the problem. Research a large oil spill and the methods used to clean it up. Prepare a presentation for your class to explain some techniques used to clean up an oil spill.

4. Plastics are synthetic polymers. Plastic soda bottles and other types of plastics can be recycled and made into other objects. Investigate how plastics are recycled and write a brief explanation of this process. Include as much of the information you know about hydrocarbons as you can.

5. Much health information centers around triglycerides, saturated fats, and unsaturated fats. Investigate these compounds and explain in a paragraph how each effects the body.

Write the letter of the functional group on the right next to the number of the compound on the left which contains that functional group.

_____ 1. ketone   a. -CHO

_____ 2. aldehyde   b. -OH

_____ 3. alcohol   c. -COOH

_____ 4. ether   d. -C=O

_____ 5. organic acid   e. -COC-

Complete the sentences by writing the correct word in the blank.

6. When methane burns it combines with oxygen to produce carbon dioxide and water. This is called a(n) _____ reaction.

7. In a compound, a double bond is broken and two bromine atoms are added to the compound. This is an example of a(n) _____ reaction.

8. Water can be removed from two molecules of ethyl alcohol to form an ether. This is an example of a(n) _____ reaction.

9. In an alkyne, a(n) _____ bond is formed between neighboring carbon atoms.

10. A(n) _____ is a compound formed by bonding together amino acids.

Circle the letter of the correct answer.

11. RNA and DNA are compounds called
    (a) amino acids.   (b) proteins.   (c) nucleic acids.   (d) peptides.

12. A triglyceride is composed of
    (a) fatty acids and glucose.   (b) fatty acids and glycerol.   (c) starch molecules.
    (d) oils and waxes.

13. A long chain of simple sugars is called a (an)
    (a) polysaccharide.   (b) lipid.   (c) nucleic acid.   (d) ether.

14. The function of an enzyme is to
    (a) synthesize a protein.   (b) form peptide bonds.   (c) act as a functional group.
    (d) speed a chemical reaction.

Answer one of the two questions.

15. Draw diagrams showing the differences among alkanes, alkenes, and alkynes.

16. Name the different chemical reactions that take place in organic compounds. Explain what happens in each reaction.

Glossary/Index

A

absolute zero   in theory, the lowest possible temperature at which all molecular motion stops   *23*

activated complex   temporary group of atoms that form as reactant particles rearrange to form products   *188*

activation energy   energy needed to move reactants into the activated complex   *188*

addition reaction   reaction in which one or more atoms or functional groups are added to an alkene or alkyne   *229*

alcohol   hydrocarbon containing the functional group -OH   *224*

aldehyde   hydrocarbon containing the functional group -CHO   *225*

alkane   hydrocarbon with only single bonds between neighboring carbon atoms   *221*

alkene   hydrocarbon with a double bond between two neighboring carbon atoms   *221*

alkyne   hydrocarbon with a triple bond between two neighboring carbon atoms   *221*

alpha particle   nucleus of a helium atom   *65*

anion   negative ion formed when a neutral atom gains electrons to have a filled outer orbit   *93*

Arrhenius acid   substance that produces hydrogen ions when it is in water solution   *159*

Arrhenius base   substance that produces hydroxide ions when it is in water solution   *159*

atomic mass unit   a unit of mass equal to 1/12 of the mass of a carbon-12 atom   *62*

atomic number   number of protons in each atom of the element   *60*

atomic particle   particle found inside an atom   *54*

Avogadro's hypothesis   equal volumes of all gases at the same temperature and pressure contain the same number of molecules   *137*

B

battery   two or more electrochemical cells connected together   *212*

beta particle   electron   *65*

bond angle   angle between two covalent bonds   *100*

Boyle's law   volume of a gas decreases as pressure increases if temperature remains constant   *34*

Bronsted-Lowry acid   a proton donor   *159*

Bronsted-Lowry base   a proton acceptor   *159*

C

calorie   amount of heat needed to raise the temperature of 1 gram of water 1 degree Celsius   *23*

carbohydrates   aldehydes or ketones with many —OH groups attached   *232*

catalyst   substance that speeds up a reaction without itself being permanently changed   *177*

cation   positive ion formed when a neutral atom loses electrons to have a filled outer orbit   *93*

Celsius scale   temperature scale in which water freezes at 0 degrees and boils at 100 degrees   *23*

Charles's law   volume of a gas increases as temperature increases if pressure remains constant   *35*

chemical bond   force that holds two atoms together   *92*

chemical change   kind of change in matter in which new substances are formed   *4*

chemical energy   energy in food, fuel, and other compounds   *15*

chemical equilibrium   state in which the forward and reverse reactions of a reversible reaction proceed at the same rate; concentration of each substance in the equation remains the same   *181*

chemical formula   group of symbols used to show the kinds of atoms and the number of atoms of each kind in a compound   *93*

chemical property   property that describes how a substance reacts with another substance   *3*

electron   atomic particle with a negative charge   *54*

electron configuration   description of the arrangement of electrons in an atom   *79*

electron orbital   a region in space where an electron is most likely to be found   *75*

electroplating   use of electric current to coat the surface of one metal with another metal   *213*

element   kind of pure substance that is the simplest kind of matter and cannot be broken down into other substances   *7*

empirical formula   formula that shows the simplest possible whole-number subscripts in the formula of a compound   *121*

end point   point in titration at which chemically equivalent amounts of acid and base are present   *169*

endothermic reaction   chemical change during which energy must be supplied   *18, 186*

energy   ability to do work   *14*

enthalpy, or H   heat given off or absorbed during a chemical reaction   *189*

entropy, or S   measure of randomness or disorder   *191*

enzyme   protein that speeds up a chemical reaction   *234*

equilibrium constant, or K_e   ratio of the concentrations of the products to the reactants in a reversible reaction at equilibrium   *181*

equilibrium stresses   changes that alter the reaction rate, adjusting the direction of reaction movement   *186*

ether   hydrocarbon containing the functional group C-O-C   *225*

evaporation   escape of vapor at the surface of a liquid   *41*

excited state   atom when one or more of its electrons are in a higher level than the lowest possible level   *71*

exothermic reaction   chemical change during which energy is released   *18, 186*

F

formula mass sum of the atomic masses of all atoms indicated in the formula of a substance   *117*

formula of an ionic compound chemical formula that tells the kinds and relative numbers of atoms of each kind in an ionic compound   *104*

free energy, or G indication of how spontaneous a reaction is, as determined by the effects of heat, temperature, and entropy   *191*

functional group group of atoms bonded together that give hydrocarbons various chemical properties   *224*

G

gamma rays very penetrating radiation like X rays   *65*

gas matter having no definite volume or shape   *30*

gram formula mass formula mass expressed in grams of a substance   *117*

ground state atom when all its electrons are in their lowest possible energy levels   *71*

group vertical column of elements in a periodic table   *83*

H

Haber process method of producing ammonia from hydrogen and nitrogen   *180*

half-cell metal placed in a solution of a salt of that metal   *207*

half-life time it takes for half of the nuclei in a radioactive sample to disintegrate   *66*

heat energy energy resulting from motion of particles of matter   *15*

heat of fusion amount of energy needed to change one gram of solid to liquid at its melting temperature   *44*

heat of vaporization amount of energy needed to change one gram of liquid to gas at its boiling temperature   *45*

hydrocarbon   organic compound with only carbon and hydrogen atoms   *220*

hydronium ion   a hydrated proton or H_3O^+ *159*

I

ideal gas   gas that obeys exactly all the gas laws   *31*

indicator   substance used to detect the presence of an acid or a base; acids and bases cause indicators to change color   *158*

ion   atom that develops an electric charge when it loses or gains electrons to fill its outer orbit   *93*

ionic bond   chemical bond formed when electrons are transferred from the outer orbit of one atom to the outer orbit of another atom   *93*

isotope   one of two or more kinds of atoms of the same element that differ from each other in their atomic masses   *61*

J

joule   unit of energy, including heat energy, in the metric system   *23*

K

K_a   ionization constant of an acid; it shows the relative strength of acids   *164*

Kelvin scale   temperature scale in which the lowest reading is 0, the lowest temperature possible   *23*

Kelvin temperature   Celsius temperature plus 273°   *31*

ketone   hydrocarbon containing the functional group C=O   *225*

kilogram   standard unit of mass in the metric system   *11*

kinetic energy   energy of motion   *15*

kinetic theory   explanation of properties and behavior of gases   *30*

kinetics   area of chemistry concerned with rates of chemical reactions   *176*

L

law of conservation of energy   the principle that energy can change from one form to another, but the total amount of energy remains the same   *19*

Le Châtelier's principle   principle stating that if a reversible reaction at equilibrium is stressed, the equilibrium will shift to relieve the stress   *185*

light energy energy from the sun and part of the electromagnetic spectrum   *15*

lipids group of compounds including fats, oils, and waxes   *233*

liquid phase of matter having a definite volume but taking the shape of the container   *41*

liter standard unit of volume in the metric system   *11*

M

mass amount of matter in an object   *11*

mass-mass problem a problem in which mass of one substance in a reaction is given and mass of another substance is found   *132*

mass number number of protons plus number of neutrons in an atom   *61*

mass-volume problem problem in which mass of one substance in a reaction is given and volume of another is found; or volume of one is given and mass of another is found   *137*

matter anything that has mass and takes up space   *2*

mechanical energy energy related to matter in motion   *15*

meter standard unit of length in the metric system   *10*

metric system system of measurement based on decimal units for length, mass, and volume   *10*

mixture material formed of substances that do not combine chemically and keep their own properties   *7*

molal boiling point elevation constant amount a one molal solution of nonelectrolyte will raise the boiling point of a solution   *154*

molal freezing point depression constant amount a one molal solution of nonelectrolyte will lower the freezing point of a solution   *155*

molality concentration of a solution expressed in moles per kilogram of solvent   *152*

molar mass mass of one mole of an element of a compound   *118*

molar volume volume that one mole of any gas occupies at STP; its value is 22.4 liters   *138*

molarity concentration of a solution expressed in moles per liter of solution   *149*

mole number 6.02×10^{23} *118*

phenolphthalein   indicator that is colorless in the presence of an acid and red in the presence of a base   *169*

physical change   kind of change in which matter may look or behave differently but no new substance is formed   *3*

physical property   property such as color, size, and texture that you can observe about matter while it is not reacting with anything else   *3*

polar   a covalent bond in which the electron pair is pulled closer to one atom, giving the bond some electric charge   *97*

polymer   large molecule made of simple molecular units repeated many times   *230*

polymerization reaction   a reaction that produces a polymer from monomers   *230*

polysaccharide   long chain of simple sugars   *233*

potential energy   stored energy   *14*

precipitates   solids formed in double replacement reactions when two solutions of ions are mixed   *130*

pressure   force of a gas on the wall of a container   *31*

products   substance or substances that are produced during a chemical reaction   *125*

property   characteristics used to describe matter   *2*

proton   atomic particle with a positive charge   *54*

pure substance   kind of matter in which all samples of the matter have the same properties   *6*

R

radical   group of atoms of different elements that act together to form an ion   *104*

rate determining step   slowest step in a reaction   *177*

reactants   substance or substances that exist before a chemical reaction   *125*

reaction   chemical change in which substances combine and form new substances   *18*

reaction rate   change of concentration in a unit of time   *176*

redox reaction   short term for an oxidation-reduction reaction   *203*

reducing agent   agent that gives electrons to another substance or loses electrons   *199*

reduction   the gain of electrons   *199*

related acid-base pair   acid and base that differ by one proton   *163*

reversible reaction   reaction in which the products can react to produce the original reactants   *181*

S

salt   compound of a positive ion other than H^+ and a negative ion other than OH^- *168*

salt bridge   tube of salt connecting two half-cells   *207*

saturated solution   solution that contains the greatest amount of solute that can be dissolved at a given temperature   *149*

single replacement   reactions in which one element takes the place of another in a compound   *129*

solid   phase of matter having a definite shape and a definite volume   *40*

solubility   number of grams of solute that will dissolve in a solvent at a given temperature   *148*

solute   part of a solution that is usually present in the smaller amount   *144*

solution   mixture of substances that are evenly spread throughout each other; particles in a solution are molecular or ionic in size   *144*

solvent   part of a solution that is usually present in the larger amount   *144*

spontaneous reaction   reaction that takes place without the addition of outside energy   *191, 207*

standard solution   solution of known concentration   *169*

stoichiometry   calculation of mass and volume relationships among substances in chemical reactions   *132*

STP   standard temperature and pressure (1 atm and 273°K)   *37*

substitution reaction   reaction in which atoms or functional groups are substituted for one or more hydrogen atoms in an alkane   *229*

surface tension   apparent "skin" effect on the surface of a liquid   *46*

synthesis   reactions in which two substances combine to form one substance   *128*

T

temperature   measure of average kinetic energy of particles of an object   *23*

titration   process of finding the concentration of an unknown solution by reacting it with a standard solution   *169*

transmutation   process of one element changing into another during a nuclear reaction   *67*

transuranic element   any element with an atomic number above 92   *67*

triglyceride   animal fat composed of fatty acids and glycerol   *233*

V

vapor   gaseous phase of substance that is a solid or liquid at room temperature   *41*

vapor pressure   pressure of vapor above a liquid   *41*

voltage   measure of electric force   *207*

volume   the amount of space, in three dimensions, that matter such as gas takes up   *11, 31*

volume-volume problem   problem in which volume of one substance in a reaction is given and volume of another is found   *136*

Periodic Table of the Elements

KEY:

| 1.01 — mass number, A |
| H — symbol |
| 1 — atomic number, Z |

1.008 H 1																	4.003 He 2
6.941 Li 3	9.012 Be 4											10.81 B 5	12.01 C 6	14.01 N 7	16.00 O 8	19.00 F 9	20.18 Ne 10
22.99 Na 11	24.31 Mg 12											26.98 Al 13	28.09 Si 14	30.97 P 15	32.07 S 16	35.45 Cl 17	39.95 Ar 18
39.10 K 19	40.08 Ca 20	44.96 Sc 21	47.88 Ti 22	50.94 V 23	52.00 Cr 24	54.94 Mn 25	55.85 Fe 26	58.93 Co 27	58.69 Ni 28	63.55 Cu 29	65.39 Zn 30	69.72 Ga 31	72.59 Ge 32	74.92 As 33	78.96 Se 34	79.90 Br 35	83.80 Kr 36
85.47 Rb 37	87.62 Sr 38	88.91 Y 39	91.22 Zr 40	92.21 Nb 41	95.94 Mo 42	98.91 Tc 43	101.1 Ru 44	102.9 Rh 45	106.4 Pd 46	107.9 Ag 47	112.4 Cd 48	114.8 In 49	118.7 Sn 50	121.8 Sb 51	127.6 Te 52	126.9 I 53	131.3 Xe 54
132.9 Cs 55	137.3 Ba 56	138.9 ★La 57	178.5 Hf 72	180.9 Ta 73	183.9 W 74	186.2 Re 75	190.2 Os 76	192.2 Ir 77	195.1 Pt 78	197.0 Au 79	200.6 Hg 80	204.4 Tl 81	207.2 Pb 82	209.0 Bi 83	(210.0) Po 84	(210.0) At 85	(222.0) Rn 86
(223.0) Fr 87	226.0 Ra 88	227.0 ●Ac 89	(261) 104	(262) 105	(263) 106	(262) 107											

★ Lanthanoid Series

140.1 Ce 58	140.9 Pr 59	144.2 Nd 60	144.9 Pm 61	150.4 Sm 62	152.0 Eu 63	157.3 Gd 64	158.9 Tb 65	162.5 Dy 66	164.9 Ho 67	167.3 Er 68	168.9 Tm 69	173.0 Yb 70	175.0 Lu 71

● Actinoid Series

232.0 Th 90	231.0 Pa 91	238.0 U 92	237.0 Np 93	239.1 Pu 94	243.1 Am 95	247.1 Cm 96	247.1 Bk 97	252.1 Cf 98	252.1 Es 99	257.1 Fm 100	256.1 Md 101	259.1 No 102	260.1 Lr 103